MIND GAMES

GEORGE LANE

MIND GAMES

AMAZING MENTAL ARITHMETIC MADE EASY

metro

First published in paperback in 2004

ISBN 1 84358 141 8

British Library Cataloguing-in-Publication Data:

A catalogue record for this book is available from the
British Library.

Design by www.envydesign.co.uk

Printed in Great Britain by Creative Print and Design, Wales

1 3 5 7 9 10 8 6 4 2

Papers used by Metro Publishing are natural, recyclable products
made from wood grown in sustainable forests. The
manufacturing processes conform to the environmental
regulations of the country of origin.

ACKNOWLEDGEMENTS

This book is dedicated to my friend David Kotin, for all the times he has said 'George, you should write a book!' I would also like to thank my friend and rival, the former Mental Calculations World Champion Robert Fountain, both for his excellent suggestions regarding the layout of this book and also for bringing to light an error which I have since corrected. I would also like to thank my brother and sister – as well as my friend Greg – for giving me their honest opinions. For those of you who want to read about the incredible feats of others, Ralf Laue and Phillip J Gould's *Alternative Book of Records* (Metro) will prove both an interesting and amazing read.

CONTENTS

PART ONE ⟫⟫⟫⟫⟫⟫⟫⟫⟫⟫⟫⟫⟫

1

NUMBER CRUNCHING
IN SECONDS

You are going to learn to do high-speed calculations *in your head*. Yes, that's right – you can put away that calculator because you won't be needing it again. You'll amaze your family and friends with some astonishing tricks as you discover the excitement of numbers.

But why spend time on numbers? Because numbers have an elegance and precision that you don't find when working with words. Translating or interpreting text from one language to another, for example, is open to judgement, especially where a word is used that has no equivalent translation. But numbers are exact and the answer will always be right or wrong. Everything is black and white, there is no grey area to upset things; two plus two will always equal four!

Just think what numbers do for us each and every day. We use them for paying bills, buying things in shops or working out the scores when playing sports. Think of this: Christmas arrives and the turkey goes into the oven. But how big is the turkey? Will it fit in the oven? How long will it take to cook? Without the use of numbers, you can't answer any of these questions.

Even if you always thought you were 'no good at maths', by the end of this book you will have no trouble with numbers. You will be able to perform complex mental calculations in seconds and be sure that you have the right answer. Is it difficult? No, nothing could be simpler. Just read on.

2

STUFF YOU REALLY NEED TO KNOW

It's important to work through the book from beginning to end. If you just dip in here and there you won't understand what's going on. Each section begins with a few problems which you should attempt *after* you have read and understood the whole section.

Try to get the answers within the time limit given. Don't use any calculating aids (other than your fingers) and don't write anything down.

The answers for each section are given at the end of the same section and you will also find a small number of questions designed to test how well you have learned the methods from that section.

If you find these too difficult, the answers are also provided but make sure you read through

5

the section again and don't attempt anything new until you're sure you understand what you have read so far.

Look out for the Champion's Top Tips that are sprinkled throughout the book and highlight the main points of how the problems can be cracked.

Don't worry if you can't answer all the problems at the start of a section. They are based on groups of questions posed in the Mental Calculations World Championships. During the four events held in 2000-2003, only nine scores of 700 points or higher were achieved. The maximum score is 750 points, and the highest score on record is 741 points, set by the (now deceased) English mind-sports expert John Rickard. If you *can* give the correct answers to all of the problems from all of the sections, then you don't need this book!

PART TWO »»»»»»»»»»

3

WORKING WITH WHOLE NUMBERS

A BIT ABOUT TERMINOLOGY

- *A sum of two numbers each of two digits will be referred to as a 'two plus two'.*
- *A sum of three numbers each of three digits is a 'three plus three plus three', and so on.*
- *A problem of subtracting a four-digit number from a five-digit one is a 'five minus four'.*
- *Multiplying two numbers of three digits each is a 'three by three'.*
- *A division of an eight-digit number by a five-digit number is an 'eight over five'. Other lengths of numbers will be similarly shown.*
- *In the text you will come across the word 'integer', which is the term we use in maths to refer to whole numbers.*

4

LIGHTNING-FAST ADDITION

Integers have no hidden 'little bits and pieces' waiting to catch you out. You'll soon see that integer additions are no harder than ordinary counting. Here is the first test section. You have 11 minutes to solve these problems (they are worth more points the harder they get). Remember, you don't attempt the tests until *after* you have read the rest of the section.

1) 18 + 96 (1 pt)
2) 87 + 68 (1 pt)
3) 122 + 386 + 649 (2 pts)
4) 604 + 848 + 954 (2 pts)
5) 11625 + 6067 + 34658 (3 pts)
6) 477632 + 26726 + 120983 + 56890 (5 pts)
7) 2060735 + 738205 + 5091372 + 827260 (6 pts)

8) 6279789 + 671069 + 8609630 + 107985 + 4205863
   (8 pts)
9) 71340690 + 23659170 + 63448968 + 3053912 +
   85936643 (10 pts)

The key to working with integers is to break each problem down into manageable steps. You can probably add a short series of two- or three-digit numbers in your head, but let's start by breaking down even these simple numbers into single digits.

HERE'S THE PROBLEM: 61 + 32 (a 'two plus two' problem) THIS IS WHAT YOU DO:

You can do this in three simple steps:

- Start by adding the last digit of each number, in this case 1 + 2. 1 + 2 = 3, so the last digit of the answer is a 3.
- Next, add the last-but-one digit of each number, i.e,. 6 + 3.
- 6 + 3 = 9, so the 'next digit up' is a 9.
- There are no more digits to add, so there are no more to find. The last step is to line the numbers up – the 3 at the end, then the 9 in front. The answer is 93.

Let's try that again with a slightly more complicated sum:

HERE'S THE PROBLEM: 99 + 89
THIS IS WHAT YOU DO:

- Again, start by adding the final digit of each number, 9 + 9. 9 + 9 = 18, but the last digit of the answer cannot be 18 as this number has two digits, not just one. In this case, we take the last digit only (the 8) and take the rest of the number (the 1) to add to the next stage of the sum. This number is known as a 'carry-over'.

- Next, we add the last but one digit of each of the numbers as before, but this time we have a carry-over; so we add that on as well – and we get 9 + 8 + 1.

  9 + 8 + 1 = 18. Again this is more than one digit, so we take only the last (the 8) and leave the rest (the 1) as a carry-over.

- The numbers in the sample problem give us no more digits to add together, so we are left with only the carry-over (the 1).

- The first digit we calculated was an 8, so the

last digit of the solution is an 8; so the part-made answer currently looks like this: ------8.

- The next digit to be discovered was also an 8, so another 8 is added to the left of the one we already have. The answer being built now looks like ----88.

- The next digit was the last to be found, a 1. Placing this to the left of the figures we already have thus gives us the final answer, 188.

Now that we've seen how easy it is to add a 'two plus two' problem, let's use some two-plus-twos to solve bigger problems.

HERE'S THE PROBLEM: 21214 + 9529 + 39610
THIS IS WHAT YOU DO:

- Here we begin by breaking down each number into pairs of digits, *starting from the right-hand end*.
- The numbers then look like this:

21214	is	2	12	14
9529	is		95	29
39610	is	3	96	10

Notice how the 'broken' numbers are lined up at the right, not the left.

- As before, the adding up starts at the right-hand end of the numbers. 14 + 29 = 43, and to add the 10 from the third number we have 43 (from the first two numbers) + 10 = 53.

- Since this number is only two digits in length, this can be placed at the end of our answer without using a carry-over. Incidentally, did you notice that we just turned a 'two plus two' problem into a 'two plus two plus two'? Easy, wasn't it...

- For now, the answer looks like -----53.

- The next step is to add the penultimate pairs of figures, 12 + 95 + 96 (another two plus two plus two). The answer to this section of the problem is 203. Consequently, the 03 part goes on to the answer and the 2 is a carry-over. Since we are dealing with pairs of digits, don't forget to add 03 to the answer rather than just 3.

- The answer now looks like ---0353.

- Now to add the rest of the figures together, not forgetting the carry-over: 2 + 3 + 2 = 7, so a 7 goes at the front of the answer, leaving a 7, then 03, then 53; the answer is 70353.

CHAMPION'S TOP TIP

When adding numbers in this way, without writing the answer as you go, you will need to remember the partial answer you have already found when you move on to the next part of the problem. This is not as difficult as you might think. Looking back at our sample problem, the first part of the problem gave the digits '----53'. Think of this as 53 in its own right, then begin adding the next set of numbers. First you see '12 with 53', then '12 + 95 = 107 with 53', and finally '107 + 96 = 203 with 53'. By thinking of the 'with 53' at each stage, you should still be able to add the numbers together and remember the 53 at the same time. Taking this to the last stage of this problem, you would begin with '2 with 0353' and finish with '7 with 0353'. It can be a little tricky at first, but minimal practice is all it takes.

You may already feel confident that you can add numbers together using a system based on breaking numbers into sets of three digits. If not, read on – I'm about to show you how simple it can be. After all, it's only a short step – how hard can that be?

HERE'S THE PROBLEM: 729652 + 91326 + 186755 + 71507

THIS IS WHAT YOU DO:

- Again, the first step is to break the numbers down into manageable pieces, this time into groups of three digits (remembering to start from the right-hand end).

  729652 becomes   729   652.

  91326   becomes   91   326.

  186755 becomes   186   755.

  71507   becomes   71   507.

- Let's just break down the last group of digits in each case; we then get: 6 52, 3 26, 7 55, and 5 07.

- As with the previous sample problem, a two-plus-two-plus-two sum should now be easy – and there should not be any real difficulty with adding more two-digit numbers on as well. Here goes for a two-plus-two-plus-two-plus-two:

- 52 + 26 + 55 + 07 = 140 (52 + 26 = 78, 78 + 55 = 133, 133 + 07 = 140). As before, the last two digits form part of the answer and the rest is a carry-over. Consequently, the partial answer is ---40 with a carry-over of

1. Now to add the single-digit parts of the numbers, not forgetting that carry-over.

- 6 + 3 + 7 + 5 + 1 = 22. The last digit is a 2, so this goes on to the answer – leaving it reading as 240. The rest of the result of the last addition is also a 2, so this is the final carry-over from the first three-plus-three-plus-three-plus-three addition completed as part of the latest sample problem.

- Using the same process for the next three-plus-three-plus-three-plus-three addition, it can be found that 729 + 91 + 186 + 71 = 1077. On adding the carry-over from the first part, the 2, we get 1077 + 2 = 1079. The '1' does not need to be carried over in this case, since there are no other digits to be added. The final answer to this sample problem is therefore the sequence 1079 followed by 240, i.e., 1079240.

How do you know when your answer is correct? Here's a method that, while not perfect, should pick up around 90% of errors.

HERE'S THE PROBLEM: 3842139 + 655697 + 1948284 + 100798

THIS IS WHAT YOU DO:

- Let's go through this one a little more quickly than before, using the techniques we've already seen.

  3842139 =   3 842 139
  655697 =   655 697
  1948284 =   1 948 284
  100798 =   100 798

- 139 + 697 + 284 + 798 = 1918; use 918 and carry over 1.
- 842 + 655 + 948 + 100 + 1 (carried over) = 2546, use 546 and carry over 2.
- 3 + 1 + 2 (carried over) = 6.
- Final result = 6546918.

There are three steps to checking the answer in a problem like this one.

STEP 1

- Check the lengths of the numbers, find the shortest, take one digit away, then drop the resulting number of figures from the end of each number. In this case, the shortest numbers are six digits long, and so five figures should be 'lost' from each number, *including the answer*. Initially, this gives us 38 + 6 + 19 + 1 = 65.

- Actually, 38 + 6 + 19 + 1 = 64. This makes the answer look wrong at first, but this is due to the 'carry-overs' of the last five digits of each of the numbers. Instead, it should be considered that the calculated answer (in this case 64) might be increased by as much as one unit short of the number of numbers involved. Here there are four numbers, 4 – 1 = 3, and 64 + 3 = 67. Thus the range available for the 'shortened answer' is from 64 to 67. Since the result is 65, it is inside the range – and the result passes the first test. Incidentally, using this step – and remembering the number of digits temporarily rejected – can help you to find an *approximate* answer to the sum.

STEP 2

- Take the last two figures (or three, if you wish) of each number, add these together and check the result against the last two or three digits of your answer. By adding the numbers in reverse order, you should be able to side-step any errors you may have made during the initial addition.

- Here we have 98 + 84 + 97 + 39 = 318, last two digits 18, matching the answer. Alternatively, using three figures at a time, we see 798 + 284 + 697 + 139 = 1918, last three digits 918, again matching the answer. Again, the answer passes the test.

STEP 3

- This is where most of the errors get 'trapped'; eight out of nine errors in addition will be caught here. This step relies on 'markers', also known as 'essences'.

- Add the values of the individual digits of a number, then check the result. If it has more than one digit, add these together and, if necessary, continue further until you have a one-digit number. This is the marker of the original number.

- Example: The first number in this addition is 3842139. 3 + 8 + 4 + 2 + 1 + 3 + 9 = 30, and 3 + 0 = 3. The marker of the number is thus

3. Using the same process, the markers of the other three numbers can be identified as 2, 9 and 7. Markers behave in the same way as the numbers they represent, thus a number with a marker of 2 plus a number with a marker of 5 will always produce a number with a marker of 7, since 2 + 5 = 7. If the result of such an addition has more than one digit, simply find the marker of that result.

- In this sample addition problem, the markers of the four numbers are 3, 2, 9 and 7. 3 + 2 + 9 + 7 = 21, and 2 + 1 = 3. The correct answer will therefore have a marker of 3. The answer arrived at was 6546918: 6 + 5 + 4 + 6 + 9 + 1 + 8 = 39, 3 + 9 = 12, and 1 + 2 = 3. This matches expectations, and so again the answer passes the test. If an answer fails *any* part of the test, then it is *definitely* wrong; passing the test completely is a *very* strong indication that it is correct.

DO YOU UNDERSTAND?

As promised, here are three questions about methods discussed in the integer additions

— A NEAT LITTLE TRICK —

Try asking somebody to gve you a series of two- (or three-) figure numbers. Using the techniques given in this section, you should be able to add each number on to a 'running total' as the numbers come in – and you should certainly have plenty of time for this if the other person is adding them up on a calculator at the same time. Just as they press the 'equals' button at the end of the list, you can give them the answer, perhaps even before they've seen it for themselves!

section. The answers are on the next page, along with the answers to the earlier nine test problems. Make sure that you understand this section completely before you go on to the next.

1) If you were using three-digit blocks, what number would you need to carry over during the addition of 93320 + 3512 + 78089 + 3126?

2) Which of the following is a *bad* way to break up the number 20802123? a) 208 021 23 or b) 20 802 123 ?

3) What is the marker of the number 5393772?

ANSWERS TO THE TIMED TEST QUESTIONS

1) 114 (1 pt)
2) 155 (1 pt)
3) 1157 (2 pts)
4) 2406 (2 pts)
5) 52350 (3 pts)
6) 682231 (5 pts)
7) 8717572 (6 pts)
8) 19874336 (8 pts)
9) 247439383 (10 pts)

ANSWERS TO 'DO YOU UNDERSTAND?'

1) The carry-over is 1, since $320 + 512 + 089 + 126 = \underline{1}047$.

2) a) is bad, since the 'broken' number should have blocks of equal numbers of digits as counted from the right-hand end. The last block in a) has two figures whereas the others each have three. These figures are broken from the left-hand end, which can lead to confusion and a wrong answer to a sum.

3) The marker is 9. $5 + 3 + 9 + 3 + 7 + 7 + 2 = 36$, and $3 + 6 = 9$.

Add up how many points you have scored for this section, remembering only to count answers which are *exactly* correct; you score nothing for an answer which is almost right.

Keep a note of the scores you achieve for each section, as the final part of the book has a table with which you can check your overall total. There is also a full-length test of 140 questions with a time limit of four hours. By the time you finish the book you should be able to tackle this test with confidence and see how it feels to take part in the Mental Calculations World Championship. With luck, your score for this test should also be higher than your total for the test questions within the individual sections, thus indicating you have improved as you have worked through this book.

5

SPEEDY SUBTRACTIONS

These problems are a little trickier than the additions, since they involve an element of 'working backwards'. This should not present too much of a problem because, just as additions are an extension of counting forwards, subtractions are an extension of counting backwards, hardly a major challenge.

Here are some test questions. Your time limit is ten minutes and, as before, the harder the problem the more marks it is worth.

1) 117 – 65 (1 pt)
2) 162 – 65 (1 pt)
3) 3677 – 1968 (2 pts)
4) 92682 – 53359 (3 pts)
5) 1227951 – 278068 (4 pts)

6) 95724476 – 77649500 (6 pts)

7) 874266364 – 750756879 (8 pts)

8) 9869732885 – 926742800 (10 pts)

Again we can start by breaking a simple problem into even smaller, simpler units.

HERE'S THE PROBLEM: 50 – 32
THIS IS WHAT YOU DO:

- Start, as with the first addition problem, by taking the final digit of each number, i.e., 0 – 2. Since this would give a negative result, which is not acceptable, the result requires a 'boost' by adding a 1 before the first part (the 0). This also results in 1 unit being taken away from the other figure of this number, the 5, thus leaving a 4. So, the last digit is 10 – 2 = 8, and this is preceded by 4 – 3 = 1, making a final answer of 18.

- In the event that you need to take 1 away from a 0 for the same reason that the 5 was reduced to a 4 here, then take 1 away from the digit before the 0. This will allow you to treat the 0 as a 10, and to take 1 away to leave 9.

- With very little practice, if indeed any is

needed at all, these 'two minus two' problems should be easy to solve directly, without breaking the numbers into single figures. Larger problems can then be broken into two-minus-twos, and then three-minus-threes. You might even try breaking some of the biggest into four-minus-fours in order to have fewer steps to complete and so save time, but the time saved would probably be outweighed by the extra time needed because of the size of the blocks you have broken your numbers into. If you *can* save time by using four-minus-fours rather than three-minus-threes, you are indeed doing well – I tend to use three-minus-threes.

HERE'S THE PROBLEM: 194019 – 108059
THIS IS WHAT YOU DO:

- Breaking these numbers into pairs of digits, in order to leave a short series of 'two minus two' problems, we get the following:
  194019   becomes   19   40   19 and
  108059   becomes   10   80   59.
- The first two-minus-two problem comes from the right-hand end. 19 – 59 gives a negative

result, and so again a 'boost' of 1 is needed. This is placed in front of the 19, and so the 40 (the previous pair of digits) is reduced to 39 (you may have noticed that we are 'borrowing' from a 0, as mentioned earlier). The first two-minus-two problem then becomes 119 – 59, giving 60, the last two digits of the answer. (If this were a problem from one of the tests, you would of course be allowed to write this part of the answer down before continuing on to find the rest of it. After all, you would not be breaking any of the rules, you would just be writing down the answer bit by bit – and there is nothing against that, not even in the World Championship itself.)

- The answer now looks like ------60, and the next pair of digits can be found by the next two-minus-two problem, which is 39-80 (remember, the 40 was reduced to 39).

- Oh dear, it seems another extra 1 is needed. The two-minus-two problem becomes another three-minus-two. I.e. 139 – 80, and the 19 preceding what was the 39 must be reduced to 18. 139 – 80 = 59, and so the answer now appears as ----5960.

- The last two-minus-two problem is 18 – 10, giving 8. The final answer is thus 85960.

CHAMPION'S TOP TIP

As with integer additions, the main problem with subtractions is remembering the partial answer as you work on the rest of the problem. The solution to this is also much the same: working through the sample problem above, we start by finding 'sixty' as a conclusion to the answer. After this comes '139 (with 60) – 80 = 59 (with 60)', i.e. 5960. The final part would therefore be '18 (with 5960) – 10 = 8 (with 5960)'. The sequence '8, 59, 60' then gives the answer (85960).

HERE'S THE PROBLEM: 148616110 – 55830341
THIS IS WHAT YOU DO:

Now it's time to see a problem broken into a series of three-minus-three problems. Having worked through one-minus-ones and two-minus-twos, it should not be too hard.

- 148616110 becomes 148   616   110. 55830341 becomes 55   830   341.
- Yet again a 'boost' is required. 110 becomes 1110, and 616 becomes 615. 1110 – 341 = 769, so the answer starts out looking like --------769.

- Next we have 615 - 830, calling for another boost. 615 becomes 1615, and 148 is reduced to 147. 1615 – 830 = 785, and the answer builds up to ----785769.
- The final part of the problem is 147-55, giving 92. The answer is therefore 92785769.

Sometimes a subtraction problem will be set to take the larger number away from the smaller one. In such a case, simply swap the numbers around, complete the problem as normal, and precede the answer with a minus sign.

CHECKING YOUR ANSWERS

As you may have guessed, steps similar to those used with integer additions can be employed to check the accuracy of answers to integer subtraction problems. The easiest way to do this is to turn the problem around, read it from right to left (starting with your answer) and treat it as an addition problem.

EXAMPLE: 747108 – 316687 = 430421

- Turning this around for checking gives us 430421 + 316687 = 747108.

- Using step 1 from the checking process in the integer additions section, we get $4 + 3 = 7$. For a more accurate check, you can use additional digits – such as $43 + 31 = 74$.

- Step 2 gives $21 + 87 = 108$, i.e., the last two figures should be 08, which they are.

- Step 3 checks the markers of the numbers. $4 + 3 + 0 + 4 + 2 + 1 = 14$, and $1 + 4 = 5$. $3 + 1 + 6 + 6 + 8 + 7 = 31$, and $3 + 1 = 4$.

- So the markers of the two numbers being 'added' are 5 and 4, and $5 + 4 = 9$; therefore the marker of the 'solution' should also be 9.

- $7 + 4 + 7 + 1 + 0 + 8 = 27$, and $2 + 7 = 9$. The marker is 9, and the answer passes the test.

- Again, failing any part of the test shows that an answer must be wrong – but please remember to check again just to be sure. You don't want to throw away an answer because just possibly you may have made a mistake in your checking! If you have checked your answer and it has definitely failed, another answer will be needed. Don't worry too much if you make a mistake – *everybody* does it.

A NEAT LITTLE TRICK

Ask someone to give you a subtraction problem: two numbers, each of six figures, with the first number being higher than the second. By breaking these numbers into blocks of two or three figures and using the other techniques in this section, you should be able to find the answer with ease and at speed.

ANSWERS TO THE TIMED PROBLEMS

1) 52 (1 pt)
2) 97 (1 pt)
3) 1709 (2 pts)
4) 39323 (3 pts)
5) 949883 (4 pts)
6) 18074976 (6 pts)
7) 123509485 (8 pts)
8) 8942990085 (10 pts)

6

MULTIPLICATION AT THE SPEED OF LIGHT (WELL, ALMOST)

Now it gets just a bit harder, but it's well worth the effort. Imagine receiving a crate of cartons of orange juice, for example – what's the best way to count them? You could count the cartons one by one, but it is much quicker to see how many cartons stand along one side of the crate and how many are across one end. Multiply the two numbers together and there, staring right at you, is your answer. Now think of a much larger crate of much smaller items; do you really want to count them all individually? I thought not...

Here are the test problems. The time limit is 19 minutes.

1) 4 x 33 (1 pt)
2) 80 x 63 (2 pts)
3) 536 x 1965 (4 pts)
4) 4112 x 1340 (5 pts)
5) 70448 x 19136 (7 pts)
6) 57485 x 264027 (8 pts)
7) 232954 x 107989 (9 pts)
8) 159061 x 1895417 (10 pts)

Here it is *definitely* advisable to start with a simple problem broken down into even simpler steps.

HERE'S THE PROBLEM: 5 X 16 (a 'one by two' problem)
THIS IS WHAT YOU DO:

- The first number (the 5) has only one digit and so does not cause any difficulties. Only the second number (the 16) can be broken down any further than it already is.
- Take the final digit of the 16, i.e.; 6, and multiply by the 5, giving 30. As with problems in the previous sections, the '0' becomes part of the answer and the '3' is a carry-over. Next we multiply 5 by 1, giving 5

(since anything multiplied by 1 gives itself) and add the carried-over 3 to give 8. This 8 is then placed before the 0 found earlier to give a final answer of 80.

Unfortunately, very few multiplications are so simple – it's time to delve a bit deeper where you will learn some new techniques. In this section, the individual digits of numbers will be referred to by the use of letters, starting with 'a' at the right-hand end. As an example, the 'a' digit of the number 3126 is the 6, as this is the last digit. The 'b' digit is 2, the 'c' digit is 1 and the 'd' digit is 3.

HERE'S THE PROBLEM: 228 X 3670
THIS IS WHAT YOU DO:

- At first, this may look like a three-by-four problem, but it is in fact a three-by-three. Any zeros on the end of a number (known as 'trailing zeros') can be written as the last part of the answer and then discarded from the main numbers; in this case we have to find the answer to 228 x 367 and then follow this with a single 0 to complete the answer to the sample problem.

- Each of the two numbers now has only three digits. Multiplication is done as a series of one-by-one problems, making it easier to remember those numbers which need to be remembered. Now it's time to work out what those one-by-ones are. To differentiate between the digits of the two numbers, the letters representing the digits of the first number will have a '1' while those for the second number will have a '2' (i.e.; for a three-by-three problem, the numbers would be c1 b1 a1 x c2 b2 a2).

- The first one-by-one problem is to multiply the two a numbers together, i.e., a1 x a2, or 8 x 7. 8 x 7 = 56, so the 6 marks the last digit of the answer (apart from the 0) and the 5 becomes a carry-over.

- Now it gets a little trickier. The second step is to multiply b1 from the first number (2) by a2 from the second (7), as well as multiplying a1 from the first (8) by b2 from the second (6). Not forgetting to add the 5 carried over from the first stage. 2 x 7 = 14, 8 x 6 = 48, and 14 + 48 = 62. Add the carry-over of 5 and the result is 67. This means that the answer so far appears as -------760, and there is a carry-over of 6.

- The one-by-one problems needed for the next digit are: c_1 x a_2 (2 x 7), b_1 x b_2 (2 x 6), and a_1 x c_2 (8 x 3). Taking these in turn: 2 x 7 = 14, 2 x 6 = 12, and 8 x 3 = 24. Adding these together, 14 + 12 + 24 = 50, and the carry-over of 6 leaves 56. Consequently, the 6 is written into the answer and the 5 becomes a carry-over.

- You should by now have realised that, regarding the first three-figure number, the first single digit to be used has been taken from one position further to the left each time; also that the final digit used from the second number has similarly been 'moving' to the left. When the 'end of the line' is reached, start taking the digits from the b_2 digit of the second number – then c_2, etc.

- The answer to this sample problem so far appears as -----6760, with a carry-over of 5. The next digit to write into the answer comes from two one-by-one problems, c_1 x b_2 (2 x 6) and b_1 x c_2 (2 x 3). 2 x 6 = 12, 2 x 3 = 6, and 12 + 6 = 18. Adding the carry-over of 5 makes a total of 23, so the next digit in the answer is 3 and the carry-over is 2. Now the answer looks like ---36760.

- Finally, we have c1 x c2 (2 x 3). 2 x 3 = 6, add the carried-over 2 to give 8, and this is the last digit to find, i.e., it is the first digit of the answer. The final answer is thus 836760.

Some multiplications involve numbers of unequal lengths. When this happens, the number of one-by-one problems used to find the digits of the answer is limited by the length of the shorter of the two numbers. Here are two examples, given in general terms, and each giving the 'lists' of one-by-one problems needed.

EXAMPLE: FOUR BY TWO

The four-figure number is d1c1b1a1, the two-figure number is b2a2. The one-by-ones are:

- First digit to be found: a1 x a2 (this is of course the *last* digit of the answer)
- Next digit: b1 x a2 and a1 x b2
- Next digit: c1 x a2 and b1 x b2
- Next digit: d1 x a2 and c1 x b2
- Last digit: d1 x b2 only

- Don't forget to add in any carry-overs you may make.

EXAMPLE: SIX BY THREE

The six-figure number is f1e1d1c1b1a1, the three-figure number is c2b2a2.

Here are the one-by-ones:

- First digit to find is as always a1 x a2 only.
- Next digit: a1 x b2 and b1 x a2
- Next digit: a1 x c2, b1 x b2, and c1 x a2
- Next digit: b1 x c2, c1 x b2, and d1 x a2
- Next digit: c1 x c2, d1 x b2, and e1 x a2
- Next digit: d1 x c2, e1 x b2, and f1 x a2
- Next digit: e1 x c2, and f1 x b2
- and finally: f1x c2 only.

By looking at the patterns of the figures used in these examples, it should be easy to work out which figures to use in which order for any lengths of numbers.

The checking system for answers to multiplications is a little more complex than that for additions.

—— CHAMPION'S TOP TIP ——

When multiplying two numbers together in this way, it is very easy to lose your way and forget just which digits to multiply next. If this happens to you, there is a simple way to put yourself back on the right track. By remembering the partial solution as you would with an addition or a subtraction, you can count the number of figures you have already calculated – and thus you can also find where to resume your efforts. An example: with the general 'six by three' problem above, after finding (for instance) four digits, you know the next one to find will be the fifth – and so you multiply $c_1 \times c_2$, $d_1 \times b_2$, and $e_1 \times a_2$. To put it another way, you use the digits which have a total of four other digits to the right of them (exactly the same number of digits you have already found).

EXAMPLE: 7294 X 13283 = 96886202

- Take the first two digits of each of the two numbers and treat them as a two-by-two problem, then check the first two digits of the result against the first two digits of the answer. 72 x 13 = 936 giving 93 as the first two digits, but this does not compare favourably with 96 as the first two digits of the answer. As with the additions, the apparent error is produced by the parts of the numbers not used in the rough calculation. If you add 1 to each of the two-digit numbers used, a higher result will be reached, in this case 73 x 14 = 1022. Since 10 is obviously *not* higher than 93, it is clear that 102 is the required number. Thus, the correct answer will begin with a sequence in the range from 93 to 102. The answer given begins with 96, and so it passes this part of the test. It should also be noted at this point that the two-by-two multiplication to give a number exactly matching the given answer, i.e., 968, would produce *three* digits, not *two*. This means we would get *one excess digit*.

- Count the number of digits 'discarded' in taking the first two digits from each of the two numbers. In this case, two digits were discarded from 7294, and three from 13283 making a total of five, one more 'excess digit' from the previous step making six. This gives the number of digits in the answer which should follow the first two. In short, the answer should be eight digits long; it is, and it passes this part of the test. Combining this step with the previous one gives a method of finding an approximate answer to the problem – much as was seen in the integer additions section.

- Take the last two digits of each of the two numbers and treat them as another two-by-two problem. The last two digits should match the last two digits of the answer. 94 x 83 = 7802, the last two digits being 02, as with the given answer – which passes this part of the test also. If you wish, you can even break down the two-by-two 'mini-problem' into some one-by-ones; this way you won't even need to find the full answer to the two-by-two since you only need to find the last two digits.

- The last step of the checking process uses the markers of the numbers.
  For 7294: 7 + 2 + 9 + 4 = 22, and 2 + 2 = 4.
  For 13283: 1 + 3 + 2 + 8 + 3 = 17, and 1 + 7 = 8.
  Since the numbers are being multiplied, multiply the markers. 4 x 8 = 32, and 3 + 2 = 5; so the marker of the answer should be 5.
  9 + 6 + 8 + 8 + 6 + 2 + 0 + 2 = 41, and 4 + 1 = 5.

This means the answer passes the last part of the test and may thus now be assumed to be correct – as indeed it is.

— A NEAT LITTLE TRICK —

Ask someone to give you a pair of three-figure numbers to multiply together. You should be able to visualise these numbers, and you shouldn't be in any real danger of losing track of where you are up to during the calculation. With practice, you will soon find yourself able to answer such a problem in a matter of seconds. If you practice with some two-figure numbers to begin with, you can 'graduate' to three figures in your own time.

Now for some questions on integer multiplications. In each case, do *not* work through the whole multiplication problem as it is shown; this *should not be necessary.*

DO YOU UNDERSTAND?
1) What are the last two digits in the answer to 48406 x 57188?
2) What is the marker of the answer to 76143 x 133524?
3) How many digits are there in the answer to 252050 x 118802?
4) Within what range of values would you find the first two digits of the answer to 136683 x 2450380?

ANSWERS TO THE TIMED TEST PROBLEMS

1) 132 (1 pt)
2) 5040 (2 pts)
3) 1053240 (4 pts)
4) 5510080 (5 pts)
5) 1348092928 (7 pts)
6) 15177592095 (8 pts)
7) 25156469506 (9 pts)
8) 301486923437 (10 pts)

ANSWERS TO 'DO YOU UNDERSTAND?'

1) 28, since 06 x 88 = 528, the last two digits being 28.

2) The marker is 9. It is worth noting that if the marker of any one of the numbers in a multiplication is 9, then the marker of the answer will always also be 9.

3) There are 11 digits. 25 x 11 = 275, and 26 x 12 = 312. This 'range' involves only three-digit numbers, giving one excess digit after the first two. Since four digits have been discarded from each number, making eight in total, these eight plus the one excess digit and the first two add to make a total of eleven.

4) The range is 31 to 35. 13 x 24 = 312, giving 31, and 14 x 25 = 350, giving 35.

7

LIGHTNING-FAST DIVISION

In this section, it will be assumed that all problems have exact integer answers unless stated otherwise.

Now that you've worked through the first three sections, this one should hold no horrors for you. Moving from multiplications to divisions is really not unlike the step from additions to subtractions – and surely that didn't hurt too much!

As you should by now be expecting, this section starts with a few problems.

Time limit: 12 minutes.

1) 756 / 9 (1 pt)
2) 6400 / 80 (2 pts)
3) 548244 / 873 (4 pts)

4) 5963594 / 806 (5 pts)
5) 52925292 / 6713 (6 pts)
6) 725080980 / 10140 (8 pts)
7) 1785098007 / 15543 (9 pts)
8) 42590159808 / 244389 (10 pts)

Here, the first step is to determine how many digits are in the answer. Take the number of digits in the smaller number from the number of digits in the larger one, which gives you a basic figure. Then take the same number of digits at the start of the larger number as there are in the smaller number and compare this sequence with the smaller number itself. If the smaller number is greater in value than this 'extracted' sequence, then your basic figure is the answer; otherwise it is one higher.

EXAMPLE: 583940 / 679
583940 is six digits long, whereas 679 has three digits. 6 – 3 = 3, so the basic figure is 3.

As 679 is 3 digits long, this is compared with the first 3 digits of 583940, i.e., 583.

It is clear that 679 *is* greater than 583, so the number of digits in the solution to the problem is the basic figure (3).

EXAMPLE: 177988024 / 17341

177988024 has 9 digits, 17341 has 5. 9 – 5 = 4, so the basic figure is 4.

As 17341 has 5 digits, this is compared with the first 5 digits of 177988024, i.e., 17798. This time we see that 17341 is *not* greater than 17798, and so the basic figure needs to be increased by 1. 4 + 1 = 5, so we would expect there to be 5 digits in the answer.

The next step is to ignore (temporarily, of course) some of the digits at the end of the larger number. How many digits to ignore? As many as is possible without leaving a number smaller in value than the smaller number from the problem.

From the example of 583940 / 679: the smaller number has 3 digits, so we need either 3 or 4 digits from the front of the larger number. The first 3 will not give a high enough value, since 679 is greater than 583. Therefore, we need the first 4 digits, giving us the number 5839.

Now to find the first digit of the answer. This is given by the answer to the question 'How many times will 679 go into 5839?' Take a digit off the end of each number in order to make an estimate, i.e. 583 / 67. This looks like it lies between 8 and 9. To be sure, check the value of 8 x 679 and compare this with 5839. The use of the full-length numbers gives better accuracy; if the numbers were longer, reduced-length versions would normally suffice. 679 x 8 = 5432, so there is room for an 8. Add another 679 and we get 6111 which is too high – so 9 is not possible. The first digit is 8.

For the next digit, take 679 x 8 (5432) away from 5839, giving 407, and add on the next digit of the larger number – the first to have been temporarily ignored, in this case the 4. Now divide the resulting number, 4074, by 679 using the same process as above. 407 / 67 is 'about 6', so try 6 x 679; this gives 4074 – exactly what is required. The second digit is 6.

Continuing in this manner shows that the next digit (the last) is a 0; so the final answer is 860.

Sometimes, it is not necessary to 'go to the end of the line' by this method, which is just as well – very large numbers dealt with in this way can make it much easier to make mistakes. Dividing by numbers ending in 1, 3, 7 or 9 can be made somewhat easier by starting the division from the 'wrong' end. Here's how it works:

HERE'S THE PROBLEM: 50711602 / 7057
THIS IS WHAT YOU DO:

- This time, take a look at the *last* digit of each number, in this case 2 and 7. Question: by what number would you need to multiply 7 to make a number which ends in 2? Try counting up in 7s – 7, 14, 21, 28, 35, 42 and there it is: 6 7s are needed, so the last digit of the answer to the problem is 6.

- Now there is *one* digit in place, this is multiplied by the last *two* digits of the smaller number (i.e., 57) and comparing the last two digits of the result with those of the larger number. 57 x 6 = 342, last two digits = 42, and the last two digits of the larger number are 02. Enough needs to be added to the 4 from the '42' to leave a number ending in '02', 0 – 4 gives a negative answer, so use 10 – 4 = 6. This means 6 is needed. Question: by what number would you need to multiply 7 to make a number which ends in 6?

- Try counting up in 7s again – 7, 14, 21, 28, 35, 42, 49, 5**6** and there it is. Eight 7s are needed this time, so the next digit to write into the answer is 8. The answer currently appears as -----86.

- Using much the same process as above, we next multiply 86 (our 'answer so far') by the last *three* digits of the smaller number, i.e., 057 (57). 86 x 57 = 4902, the last three digits are **9**02. Since the last three digits of the larger number are **6**02, the 9 needs to be 'added to' to leave a 6; thus a 7 is required, as 9 + 7 = 16. Since it is clear that only one 7 is needed to make a number ending with a 7, the next digit to be written in is a 1. The answer now looks like ---186.

It is possible to continue in this manner to find the next digit the same way, but it is easier to work from the start of the answer at this point. As a rule, it is not normally advisable to try to find more than three digits in this way – the multiplications needed can be somewhat trickier than those used in working from the start of the answer. By following the first steps given earlier, the answer can be seen to be four digits long with the first digit being 7; so the final answer is 7186.

Working from both ends of the problem can seem a little like 'burning the candle at both ends', but – just like that candle – the problem is finished very quickly (almost a 'blaze of glory'!).

If the divisor (the smaller number) ends with a 5 you can double it (or the last four digits), reject the resulting final 0, and take the last three digits of what is left. By then doing the same with the larger number, you can use these two three-digit numbers in exactly the same way as before to find the last three digits of the answer.

HERE'S THE PROBLEM: 70097796965 / 276545
THIS IS WHAT YOU DO:

- The last four digits of the smaller number are 6545, doubled this gives 13090, losing the 0 gives 1309, and the last three digits are 309. Using the same process for the larger number, the number arrived at is 393. These numbers can be used as above to find the last three digits of the answer, namely 477.

- Following the first steps from before, the first *two* digits can be found to be 25; now another technique can be employed to find the remaining digit.

- At the moment, the answer is 25-477, with one missing digit. It may be no surprise to learn that the 'new' technique involves markers. The marker of the larger number is 2, and the marker of the smaller number is also 2. This time the question is: by what number (in the 1 to 9 range) would the 2 from the smaller number need to be multiplied to give a number with a marker of 2?. In this case there is no need for 'counting up' as before, since the answer is clearly 1. Consequently, the marker of the answer must also therefore be 1.

- Next, check the marker of the partial answer. The digits found so far are 2, 5, 4, 7 and 7, giving a marker of 7. To *raise* this (*not* lower it) to a marker of 1, this needs a 3 (raising from 7 to 10). This 3 is thus the missing digit, leaving a final answer of 253477.

If the marker of the smaller number is 3 or 6, this technique will only leave a set of three different digits to choose from (1-4-7, 2-5-8, or 3-6-9), meaning that some more work 'from the front' would be required – although only a rough estimate would be needed to reduce the choices to a single digit. If the marker of the smaller number is 9, this technique cannot be used at all.

━━━━━━ **CHAMPION'S TOP TIP** ━━━━━━

When the divisor is an even number, the technique of working 'from the wrong end' can still be used, although a little 'tweak' is needed. In order to use this method, you will need to check the smaller number to see how many times it can be divided by 2. Once you have reduced the smaller number to an odd number, you can reduce the larger number by the same ratio (i.e. divide both numbers by 4 or 8) to produce another odd number – and then you can work 'from the wrong end' as above. You may, if you wish, still work with the original numbers in order to find the other figures.

INTEGER DIVISIONS WITHOUT EXACT INTEGER SOLUTIONS

If an integer division problem does not have an exact integer solution, or if there is no notice that the answer is of this form, the technique of working from the end of the answer cannot be used. Instead, it is necessary to work from the front of the number 'filling in' with additional zeros when the end of the larger number is passed (not forgetting, of course, to put the decimal point in at the right place).

HERE'S THE PROBLEM: 722 / 35
THIS IS WHAT YOU DO:

- 72 / 35 gives a 2 for the first digit, and leaves a remainder of 2. The answer begins by looking like 2------.

- The 2 followed by the remaining 2 from the larger number gives us 22, but this will not support any multiple of 35, and so the next digit is 0 and the remainder is 22.

- The answer now looks like 20------.

- There are no more digits to take from the larger number; this is the cue to enter the decimal point into the answer. To continue finding digits for the answer, add a 0 to the remainder (in this case 22) and use that. 220 / 35 gives 6, 6 x 35 = 210, and 220 - 210 = 10. The answer now looks like 20.6---, and the remainder is 10. Continuing in the same manner, 100 / 35 gives 2 with a remainder of 30. This leaves the answer looking like 20.62----.

- Further calculations can be made until either the desired number of decimal places is exceeded (in order to allow for rounding) or the remainder becomes 0.

A NEAT LITTLE TRICK

Ask someone to write down a three-figure number and show it to you. Next, ask them to multiply this by a 'secret' three-figure number and show you the result. Using the techniques given in this section, you should be able to divide the larger number by the smaller one very quickly; then you can reveal the 'secret' three-figure number.

DO YOU UNDERSTAND?

Assume all problems have exact integer answers, but do *not* work through the problems in full to answer the questions. This should not be necessary in order to answer the questions.

1) How many digits are there in the answer to 77958 / 122?

2) What are the first two digits in the answer to 8601242 / 962?

3) What are the last two digits in the answer to 204768468 / 9381?

ANSWERS TO THE TIMED TEST PROBLEMS

1) 84 (1 pt)
2) 80 (2 pts)
3) 628 (4 pts)
4) 7399 (5 pts)
5) 7884 (6 pts)
6) 71507 (8 pts)
7) 114849 (9 pts)
8) 174272 (10 pts)

ANSWERS TO 'DO YOU UNDERSTAND?'

1) There are 3 digits.
2) The first two digits are 89.
3) The last two digits are 28.

Remember again to keep a record of your score; there are plenty of sections left to go before you reach your final total.

PART THREE

8

A QUICK WAY WITH DECIMALS

Here I shall be using more terminology for ease of reference to the sizes of numbers. Numbers may be referred to by quoting how many digits are present on either side of the decimal point, such as 521.97 being a 'three and two', and 27498.065 being a 'five and three'.

These problems are, as you probably expect, similar to the integer additions. However, the numbers 'spill over' beyond the decimal point – and so you need to find a common 'finishing point' before you can reach a final answer. Are you ready? Here we go...

9

SPEEDY DECIMAL ADDITIONS

I hope this is no surprise, but this section starts with a few problems.

Time limit: 15 minutes.

1) 6.8 + 3.3 (1 pt)
2) 74.6 + 30.6 (2 pts)
3) 97.6 + 48.4 (2 pts)
4) 830.31 + 5961.1 + 8912.777 (4 pts)
5) 32518.655 + 54738.936 + 3011.662 (5 pts)
6) 44388.054 + 74106.464 + 68607.858 + 66795.032 (6 pts)
7) 17799.9766 + 42590.5017 + 60337.1609 + 20277.5322 (7 pts)
8) 9494.8695 + 82351.1328 + 280611.7305 + 164238.9183 (8 pts)

9) 7972258.41701 + 188764.9224 +
258433.36163 + 850497.23215 +
5451856.46987 (10 pts)

Additions of decimal numbers can be treated in a very similar manner to integer additions. Start by adding together the parts of the numbers which follow the decimal points, making sure that you treat each such partial number as being of the same length – using trailing zeros to increase number lengths if necessary. Next, add the pre-decimal sections as more integers, not forgetting any carry-overs.

HERE'S THE PROBLEM: 22.7 + 55.1
THIS IS WHAT YOU DO:

- Here, each of the numbers is a 'two-and- one'. Consequently, the post-decimal sections are of the same length (one digit) and therefore no 'filling' with trailing zeros is needed. Treating these post-decimal numbers as a one-plus-one addition, we see this as 7 + 1 = 8. This shows that the post-decimal part of the answer is 8, and there is no carry-over.
- The pre-decimal sections of the numbers

provide us with a two-plus-two addition. 22 + 55 = 77, and so the pre-decimal part of the answer is 77. The full answer is thus 77.8.

HERE'S THE PROBLEM: 865.21 + 4407.04 + 1456.839
THIS IS WHAT YOU DO:

- Since these numbers have different post-decimal lengths, these portions of the numbers need to be treated as being of equal length. This means that for purposes of addition the three-and-two requires a trailing zero to turn it into a three-and-three (i.e. 865.210), and the four-and-two requires similar treatment (becoming 4407.040).

- Now adding the post-decimal sections as a three-plus-three-plus-three problem, we have 210 + 040 + 839 = 1089. Since the partial numbers have three digits and the partial answer has four, the initial '1' is a carry-over and the remaining '089' is the post-decimal part of the answer.

- Using some more basic additions, we find that 865 + 4407 + 1456 = 6728, and the carry-over makes it 6729. Thus the final answer is 6729.089.

—— CHAMPION'S TOP TIP ——

When decimal numbers like these are to be added together, you might like to try adding them as if they were integers (remembering, of course, how many digits there should be after the decimal point). Count how many digits are in your integer 'answer' and subtract the number of digits after the point, then you know how many digits to quote before the point. This should make it easy to put the point in the right place and thus leave you with a complete answer to the problem.

HERE'S THE PROBLEM: 9093174.21515 + 970392.2596 + 410482.39897 + 212575.90685 + 6965910.0785
THIS IS WHAT YOU DO:

- Now that larger numbers are being used, it's about time to break the parts either side of the decimal points into smaller units using the techniques in the integer additions section. As some of the numbers are 'something and four' and others are 'something and five', they must – as before – be treated as having the same post-decimal length by adding trailing zeros where necessary.

- 9093174.21515 becomes 9093 174. 21515

- 970392.2596   becomes   970   392.   25960
- 410482.39897   becomes   410   482.   39897
- 212575.90685   becomes   212   575.   90685
- 6965910.0785   becomes   6965   910.   07850
- The post-decimal part of the sum is thus 21515 + 25960 + 39897 + 90685 + 07850, and can be broken down into a two-plus-two-plus-two-plus-two-plus-two problem and a three-plus-three-plus-three-plus-three-plus-three. To begin with, we have 515 + 960 + 897 + 685 + 850. This gives 3907, and so the final three digits of the post-decimal part of the answer are 907 and the 3 is a carry-over.
- Next we have 21 + 25 + 39 + 90 + 07, giving 182. Adding the carry-over of 3 gives 185, and so the remaining post-decimal figures are 85 and the 1 is to be carried over to the pre-decimal part of the addition. The answer currently looks like ------.85907.
- Now for the pre-decimal part: starting with the addition of three-digit blocks, we have 174 + 392 + 482 + 575 + 910 = 2533 and adding the carry-over gives 2534. Thus 534 gets written into the answer and the 2 is carried over.

- The next part of the addition is 093 + 970 + 410 + 212 + 965 = 2650, and 2650 + 2 = 2652. The 2 is the last carry-over, and the 652 sequence gets written into the answer. Adding the carry-over to the last remaining digits of the problem, we get 9 + 6 + 2 = 17. Writing these digits into the answer, this finally appears as 17652534.85907.

- As with the integer additions, it is possible to check the accuracy of the answers. Simply use the pre-decimal parts of the numbers to copy step 1, use the post-decimal parts (complete with the 'extra' trailing zeros where necessary) to copy step 2, and simply add the digits as before to find the markers as in step 3.

EXAMPLE: 17315.4408 + 53963.995 + 69414.7388 + 71169.2987 = 211863.4733

As these four numbers are all five-and-three or five-and-four, the first test is 17 + 53 + 69 + 71 = 210, leaving the first three digits of the answer in the range 210 to 213. The digits showing are 211, and the answer passes the test. In step 2, we have the numbers 4408, 9950, 7388 and 2987. To check by using the last two digits, we have 08 + 50 + 88 + 87 = 233. This means the last two digits of the answer should be 33, and they are. The markers of the numbers are 6, 4, 5 and 5. As 6 + 4 + 5 + 5 = 20, and 2 + 0 = 2, we would expect the marker of the answer to be 2, which it is.

A NEAT LITTLE TRICK

Ask a friend to write down a short series of decimal numbers such as those we have just been working with. Most people would expect you to start writing your answer at the right-hand end, but here's a way you can surprise them by starting at the left. Add up the post-decimal parts of the numbers and calculate the carry-over. Bearing this number in mind, you can treat the rest of the sum as an integer addition (although you may need to add the post-decimal parts again later if you don't remember that part of the answer). Having already accounted for the carry-over from this part, you can disregard it should you need to create it a second time.

DO YOU UNDERSTAND?

1) What are the post-decimal digits in the answer to 34292.341 + 14108.462 + 5822.99?

2) What size (i.e., 'one and one') is the answer to 18.4 + 47.8?

3) What is the marker of the answer to 411.92 + 1409.72 + 3770.652?

ANSWERS TO THE TIMED TEST PROBLEMS

1)   10.1 (1 pt)

2)   105.2 (2 pts)

3)   146.0 (2 pts)

     Also acceptable is 146, without the '.0'.
     Trailing zeros after the decimal point are
     not entirely necessary, they are merely
     indicative.

4)   15704.187 (4 pts)

5)   90269.253 (5 pts)

6)   253897.408 (6 pts)

7)   141005.1714 (7 pts)

8)   56696.6511 (8 pts)

9)   14721810.40306 (10 pts)

ANSWERS TO 'DO YOU UNDERSTAND?'

1)   The digits are 793.

2)   The answer is a 'two and one' number
     (the answer is 66.2).

3)   The marker is 7.

10

SPEEDY DECIMAL SUBTRACTIONS

Now you've been through decimal additions, it's time to turn to subtractions. After all, the same step with integers wasn't too hard, was it?

Again we start with some problems to be solved.

Time limit: 9 minutes.

1) 65.6 – 26.7 (1 pt)
2) 1029.5 – 144.7 (2 pts)
3) 984.98 – 461.55 (3 pts)
4) 1766.96 – 623.05 (3 pts)
5) 1649.247 – 1106.87 (4 pts)
6) 92185.082 – 65882.009 (6 pts)
7) 343411.267 – 142355.4391 (8 pts)
8) 2406792.29556 – 990536.31219 (10 pts)

Just like the additions, decimal subtractions can be treated in a similar manner to the corresponding problems with integers. Decimal numbers in subtraction problems can be split in the same way as those in additions, in groups of two or three digits from the decimal point outwards.

HERE'S THE PROBLEM: 4103.75 – 1924.37
THIS IS WHAT YOU DO:

- Breaking the numbers into blocks of two digits, we find:
  4103.75 becomes   41   03   . 75.
  1924.37 becomes   19   24   . 37.
- Starting with the post-decimal digits, we have 75 – 37 = 38 and thus the answer to the problem begins by looking like ------.38.
- The first pairs of pre-decimal digits in the numbers (the closest to the decimal point) give 03 – 24. As with the integer subtractions, we need to 'boost' the 03 to make 103, consequently reducing the 41 to 40.
- 103 – 24 = 79, so the partial answer now shows as ---79.38.
- With the 41 reduced to 40, the final part of

the problem is 40 – 19. This gives 21, and so the final answer to the problem is 2179.38.

Larger numbers can be dealt with in much the same way, as with the following problem in which the numbers are broken into three-digit blocks.

HERE'S THE PROBLEM: 1490951.58806 – 517138.51058
THIS IS WHAT YOU DO:

- 1490951.58806 becomes 1490 951 .   588 06.
- 517138.51058   becomes   517   138 .   510 58
- The first part of the solution is found, as usual, by taking the last pair of digits from each number. 06 – 58 gives a negative answer, so another boost is required. This leaves us with 106 – 58, giving 48, and the 588 being reduced to 587. The partial answer so far is -----.---48.
- Now taking the other three post-decimal digits, the next part of the solution comes from 587 – 510. Since this gives 77, it should be noted that as the blocks of digits used have three digits each then this part of the answer should also have three digits. Thus, the answer now appears as ------.07748.

- The next part of the problem is 951 – 138, giving 813 and partial answer so far looks like ---813.07748.
- The next part is 490 – 517, which is negative, and so it's time for another boost. 490 becomes 1490, the preceding 1 is reduced to nothing and so 1490 – 517 is the final part of the problem. As this gives 973, the final answer to this problem is 973813.07748.

As with other problems, the accuracy of the answers can be checked by using a few simple tests.

EXAMPLE: 109293.051 – 22698.05 = 86595.001

- Turning this problem around, as we did in the integer subtractions section, we get 86595.001 + 22698.05 = 109293.051. Taking the post-decimal figures, we first have to add a trailing zero to the end of the second number so that it has the same number of post-decimal digits as the other numbers. This leaves us with 001 + 050 = 051, which is correct – and the answer passes this test.

CHAMPION'S TOP TIP

Decimal subtractions can be treated like integer subtractions if you take care. As well as the possibility of temporarily ignoring the decimal points, there is another way: first, check to see if the post-decimal part of the second number is larger than that of the first. If it is, imagine the pre-decimal part of the second number is larger by 1 than it really is; if not, don't bother. Now treat the two pre-decimal parts as giving an integer subtraction, and this will give you the pre-decimal part of the answer. Now treat the post-decimal parts (including any necessary trailing zeros) as another integer subtraction. Did you spot the problem? Well, the post-decimal part of the first number might be smaller than that of the second – in which case you imagine the post-decimal part of the first number as having a '1' as an extra digit at the beginning. The losing and gaining of the '1' is the same as the carry-over or the boost.

- Next, by removing the last three digits from the pre-decimal part of each of the numbers, we get 86 + 22 = 109. Actually, 86 + 22 = 108, but don't forget that as we are 'adding' two numbers we must also be prepared to add up to (2 – 1 = 1) to our 'answer' at this stage. The total should thus be in the range of 108 to 109 – which it is.

Again, the answer passes the test.

- The final step is to check the markers of the numbers. 86595.001 has a marker of 7, and 22698.05 has a marker of 5. 7 + 5 = 12, and 1 + 2 = 3, so the marker of the answer should also be 3. It is, and the answer passes the final test.

A NEAT LITTLE TRICK

Ask someone to write down two large decimal numbers for you – the larger the better, but don't try to take on very large numbers without a little practice first. Using the techniques on display in this section, you should have no difficulty in finding the difference between the two numbers. In fact, try asking for a series of ten such subtraction problems – and see the looks on other people's faces when they see how fast you are at finding all the right answers.

DO YOU UNDERSTAND?

1) What are the post-decimal digits in the answer to 118236.651 – 44910.32?
2) What 'size' is the answer to 246549.918 – 105151.4785?
3) What is the marker of the answer to 1797.555 – 1411.09?

ANSWERS TO THE TIMED TEST PROBLEMS

1) 38.9 (1 pt)
2) 884.8 (2 pts)
3) 523.43 (3 pts)
4) 1143.91 (3 pts)
5) 542.377 (4 pts)
6) 26303.073 (6 pts)
7) 201055.8279 (8 pts)
8) 1416255.98337 (10 pts)

ANSWERS TO 'DO YOU UNDERSTAND?'

1) The digits are 331.
2) The answer, 141398.4395, is a six-and-four number.
3) The marker is 5.

11

HIGH-SPEED DECIMAL MULTIPLICATIONS

I'm sure you realised this section would be next. These problems are not as taxing as you might expect; the only real difference between decimal and integer multiplications is the question of where to put the decimal point. Once that little 'point of order' is out of the way, the rest is quite simple.

Here are the test problems for this section.

Time limit: 20 minutes.

1) 6.4 x 1.8 (1 pt)
2) 83.3 x 16.4 (2 pts)
3) 422.5 x 686.9 (4 pts)
4) 53.61 x 136.11 (5 pts)
5) 641.45 x 698.39 (6 pts)
6) 942.93 x 2412.42 (7 pts)

7) 5872.31 x 9867.98 (8 pts)
8) 2187.258 x 7966.477 (10 pts)

Again, the decimal problems can be treated similarly to integer problems. When multiplying the numbers, the only problem is the correct positioning of the decimal point. To find how many post-decimal digits your answer should have, simply count how many post-decimal digits there are in the two numbers and add the figures together. Watch out for a catch, however, coming soon...

HERE'S THE PROBLEM: 2.6 X 3.6
THIS IS WHAT YOU DO:

- If the decimal points are temporarily ignored, the problem becomes 26 x 36. By using techniques already seen in the integer multiplications section, this easily gives us 936. Since each of the original numbers is a one-and-one number, there are two post-decimal digits in total. The final answer is therefore 9.36.

- When writing down the answer piece by piece in a test or competition, you should remember first to count the post-decimal digits and to write the decimal point into your answer when you have entered the appropriate number of figures.

- Sometimes a decimal multiplication gives a result ending in one or more trailing zeros. When this happens, there is no need to write any of them down unless the set of zeros extends to the left of the decimal point; then they become necessary. However, you *must* remember how many zeros you have 'rejected', since they still count towards the total of the post-decimal digits.

HERE'S THE PROBLEM: 617.76 X 8114.25
THIS IS WHAT YOU DO:

- Here we have a three-and-two number and a four-and-two, therefore there are *apparently* four post-decimal digits in the answer. As promised, here comes the catch.
- By working through the steps as shown in the integer multiplications section, the post-decimal digits (the first four to be calculated) are found to be 0800. As the trailing zeros do not need to be written

down, the answer starts by appearing as
-----.08.

- Continuing with the same steps in order to complete the multiplication, the remaining digits are revealed as 5012659, giving a final answer of 5012659.08.

CHAMPION'S TOP TIP

As I mentioned in the first sample problem, it's easy to treat the multiplication as being one of integers. With a little practice, for that is all it takes, it should be quite a simple matter to replace the decimal point in the right place. Just don't forget about those pesky little trailing zeros! Whether you finish the problem in your head or write it down figure by figure, this is by far the easiest way to beat this type of problem. And, as with previous problem types, remember the figures you have already 'created' as a single number – it's easier than it sounds.

As with previous problems, the accuracy of the answers can be checked with a few simple tests.

EXAMPLE: 29.5 X 423.43 = 12491.185

- Here we have a multiplication of a two-and-one number by a three-and-two. Multiplying the first two digits of each number, we have 29 x 42 = 1218. Adding 1 to each of these two-digit numbers gives 30 x 43 = 1290. Consequently, the first four digits of the answer should be in the range of 1218 to 1290. As the first four digits (with or without the decimal point) are 1249, the answer passes this test.

• Looking at the last two digits of each of the two numbers (disregarding the position of the decimal point), we see 95 and 43. Since 95 x 43 = 4085, we should expect the last two digits of the answer to be 85; this is indeed the case.

• As this multiplication clearly produces no post-decimal trailing zeros, we can find the number of post-decimal figures by adding the numbers of post-decimal digits in the original numbers. As 1 + 2 = 3, we should expect the answer to have three digits after the decimal point. It has.

• Now to check the number of pre-decimal digits: using the method from the integer multiplications section, we find that 29 x 423 has 5 digits. We have already seen (above) that 'dropping' one digit (the 3) from 423, we get 29 x 42 = 1218 which has 4 digits. Restoring the single 'dropped' digit, the total becomes 5.

- Checking the markers is the final test. Remember that the marker of a number disregards the position, even the presence, of a decimal point. For the first number, 2 + 9 + 5 = 16 and 1 + 6 = 7. For the second, 4 + 2 + 3 + 4 + 3 = 16 and 1 + 6 = 7. Multiplying out, we find 7 x 7 = 49, 4 + 9 = 13 and 1 + 3 = 4. We should thus expect the marker of the answer to be 4. 1 + 2 + 4 + 9 + 1 + 1 + 8 + 5 = 31, 3 + 1 = 4, the correct figure. The answer has passed all of the tests, and we can treat it as being correct – which it is.

EXAMPLE: 8683.38 X 5457.65 = 47390848.857

- This is a multiplication of two four-and-two numbers. Using the first two digits again, we get 86 x 54 = 4644. Raising each by 1 we get 87 x 55 = 4785 and so the first four digits of the answer should be in the range of 4644 to 4785. As 4739 is within that range, the answer passes the first test.

- Using the last two digits, we get 38 x 65 = 2470. This appears to give the last two digits of the answer as 70, but the '0' is a trailing zero after a decimal point. This single trailing zero means that the number of post-decimal digits should be reduced by 1. As this only gives one definite digit, the 7, it is advisable to use three digits from each number. 338 x 765 = 258570, the last three digits being 570. This shows that the last two digits should be 57, as they are, and there is still that single trailing zero to consider.

- With two post-decimal digits in each number, we should expect to see four post-decimal digits in the answer (since 2 + 2 = 4). However, we have already seen that there is a trailing zero; thus, the number of post-decimal digits is reduced by one, leaving three. This is indeed how many we see, and thus the answer passes this test also.

- Using the same process as before, we find that there should be eight digits before the decimal point. There are, and the answer passes another test.

- As the marker of one of the two numbers is

9, the marker of the answer should also be 9 – as indeed it is. The answer is correct.

— A NEAT LITTLE TRICK —

Ask somebody to give you a pair of 'two-and-one' numbers – or, if you're feeling adventurous, maybe 'two-and-two'. You've already seen how to solve problems like this one, so the answer should not be hard to find. Your friend, however, might well be left wondering…

DO YOU UNDERSTAND?

1) What 'size' is the answer to the problem 4654.55 x 2248.18?

2) What are the post-decimal digits in the answer to 28.9 x 42.9?

3) In what range would you expect to find the first four digits of the answer to 89.66 x 679.93?

4) What is the marker of the answer to 28.9 x 42.9?

ANSWERS TO THE TIMED TEST PROBLEMS

1) 11.52 (1 pt)
2) 1366.12 (2 pts)
3) 290215.25 (4 pts)
4) 7296.8571 (5 pts)
5) 447982.2655 (6 pts)
6) 2274743.1906 (7 pts)
7) 57947837.6338 (8 pts)
8) 17424740.550066 (10 pts)

ANSWERS TO 'DO YOU UNDERSTAND?'

1) This gives an 'eight-and-three' answer. Not eight-and-four, since the answer (10464266.219) carries one post-decimal trailing zero.
2) The post-decimal digits are 81.
3) 5963 to 6120.
4) The marker is 6. The markers of the original numbers are 1 and 6, and 1 x 6 = 6.

Keep recording those scores, there's quite a way to go yet...

₁₂

DECIMAL DIVISIONS AT THE DOUBLE

Decimal divisions can be tricky. The size of the answer is not always obvious, but a little practice can work wonders. The main problem is to find the number of figures after the decimal point. Here are your timed test problems for this section. All problems have exact answers.

Time limit: 18 minutes.

1) 7.22 / 3.8 (1 pt)
2) 120.69 / 8.1 (2 pts)
3) 1357.07 / 94.9 (3 pts)
4) 205729.05/ 320.7 (5 pts)
5) 55633.504 / 618.7 (6 pts)
6) 193272.561 / 959.55 (8 pts)

7) 56695208.6272 / 6955.52 (9 pts)

8) 158166.537561 / 362.419 (10 pts)

This section can be dealt with in much the same manner as integer divisions, but the position of the decimal point needs to be determined, and the number of post-decimal digits is not easy to determine.

The numbers can be treated as integers with the decimal point being 'replaced' after the answer has been found. The number of digits after the decimal point can be found by subtracting the number of post-decimal digits in the second number from those in the first. This is not always accurate. If we treat the numbers as integers, we can see that each of these numbers may have factors of 2 or 5 (that is to say that each number may be divided by 2 or 5 a number of times). If the second number has more of these factors than the first, i.e., it can be divided by 2 or 5 more times, then this number of 'extra' factors will increase the number of post-decimal digits in the answer.

The method only works for divisions with exact answers. We should only look at 'extra' factors if the second number has them – if the first number has 'extra' factors of 2 but not 5 and the second number has 'extra' factors of 5 but not 2, we should only concern ourselves with 'extra' factors of 5.

If the first number has more "extra' factors than the second, the answer should not be altered.

HERE'S THE PROBLEM: 3358.23 / 47.1
THIS IS WHAT YOU DO:

This provides a relatively simple beginning, since neither number (when treated as an integer) is divisible by either 2 or 5.

Treating the numbers as integers by disregarding the decimal points, this problem then becomes 335823 / 471. Using the methods discussed in the integer divisions section, we reach an answer of 713. Since the first number has two post-decimal digits and the second number has one, and (of course) 2 – 1 = 1, we can expect the answer to have one digit after the decimal point. This suggests the answer is actually 71.3. Taking only the integer parts of the numbers, we can see that the answer is approximately 71; thus suggesting the same answer of 71.3, which is correct.

HERE'S THE PROBLEM: 795635.05 / 453.25
THIS IS WHAT YOU DO:

This is where things aren't quite so easy to deal with. If we treat the numbers as integers, as we have before, we can see that the second number can be divided by 5 twice over; the first number can be divided by 5 only once.

This means that the second number has an 'extra' factor of 5 as mentioned previously. The numbers of post-decimal digits in the original numbers (two in each case) tend to indicate that the answer will have no digits after the decimal point (since 2 – 2 = 0). However, the single 'extra factor of 5' in the second number means that the answer should now have a single digit after the decimal point. The division *can* still be dealt with by treating the numbers as integers; this is achieved by adding a 0 on to the end of the first number.

Consequently, the problem is presented in integer form as 795635050 / 45325. By using the methods discussed in the integer divisions section, the answer is given as 17554. Knowing that there is one digit after the decimal point, this must mean that the final answer is 1755.4, which it is.

HERE'S THE PROBLEM: 46.8 / 1.95
THIS IS WHAT YOU DO:

- At first glance, the number of post-decimal digits in the answer to this problem appears to be −1. However, we now treat the numbers as integers and check for factors of 2 and 5.

- With the decimal points disregarded, the numbers are 468 and 195. It is clear that the second number can be divided by 5 once only and the first cannot be divided by 5 at all. In short, the second number has one extra factor of 5 – and so we need to add one 0 to the end of the first number before we perform the integer division. This also means that the number of digits after the decimal point is raised by 1 from −1 to 0.

- Following on from this, the integer representation of the problem is 4680 / 195. This gives an answer of 24. We have no post-decimal digits to assign, and so we must now check the size of the answer (is it 24, or 240, or an even larger answer?). Taking two numbers which roughly represent the original numbers, this gives a 'new' problem of 48 / 2. This gives an answer of 24, i.e. two digits before the decimal point, and so this is what we must look for in the answer. The answer is thus 24.

Decimal divisions with inexact answers can be dealt with in exactly the same way as inexact integer divisions, after adjusting (if necessary) the smaller post-decimal length by adding any required trailing zeros, and so there is no real reason to consider them at this point.

Showing another similarity with integer divisions, answers to decimal division problems can be checked by turning each problem around and treating it as a multiplication problem. All the steps are the same as for the 'genuine' decimal multiplications, but the process for

---

— CHAMPION'S TOP TIP —

When dividing decimal numbers, they may indeed be treated as integers. Bearing in mind the effects of having factors of 2 or 5, this might seem to confuse matters – but there is an 'escape hatch' to beat this. By looking at 'rough copies' of the original numbers (as we have just done in finding the final answer of '24' above), it is easy to see roughly what the final answer should be. This tells you how many digits to place before the decimal point, leaving any others to follow it.

checking the markers of the numbers is limited in the same manner as mentioned in the integer divisions section.

EXAMPLE: 15167.564 / 369.4 = 41.06

- Turning this problem around, it becomes 41.06 x 369.4 = 15167.564.
- Taking the first two digits of each number, we find the 'opening range' for the answer to be 1476 to 1554 (41 x 36 = 1476 and 42 x 37 = 1554). Since the first four digits of the answer are 1516, the answer passes this test.
- Next, we take the last two digits (i.e., 06 and 94). 06 x 94 = 564, and so the answer should conclude with 64. It does, and so it passes another test.
- Note that there are no trailing zeros, and so there is no adjustment required to the number of post-decimal digits in the answer. Since the first number has one such digit and the second has two, the answer should have a total of three – which it does.
- Using the same method as described in the decimal multiplications section, we can see that this multiplication of a two-and-two

number by a three-and-one number should give an answer with five digits before the decimal point. This is indeed what we see, and so the answer passes another test.

- Now to check the markers: the marker of 41.06 is 2, and that of 369.4 is 4. Since 2 x 4 = 8, the answer should have a marker of 8. This is the case, and so the answer has passed all of the tests and can be regarded as correct – which it is.

- It is important to remember that if the divisor (the second number) in the original division problem has a marker of 3 or 6 then the usefulness of this step is diminished. Also, if the divisor has a marker of 9 then it will be useless and should *not* then be used as it would be a waste of valuable time.

A NEAT LITTLE TRICK

Ask a friend to write down a 'two-and-two' number and keep it secret. Then they should write down a similar number on another piece of paper, along with the result of the two numbers multiplied together. Using the techniques from this section, you should be able to divide the large number by the small one and tell your friend the 'secret' number rather quickly. After all, you already know the size of the number you are looking for!

DO YOU UNDERSTAND?

1) What 'size' is the answer to the problem 420138.2384 / 465.62?

2) What are the last two digits in the answer to 5582.56 / 8.51?

3) What are the first two digits in the answer to 420138.2384 / 465.62?

4) What is the marker of the answer to 27.52 / 3.2?

ANSWERS TO THE TIMED TEST PROBLEMS

1) 1.9 (1 pt)
2) 14.9 (2 pts)
3) 14.3 (3pts)
4) 641.5 (5 pts)
5) 89.92 (6 pts)
6) 201.42 (8 pts)
7) 8151.11 (9 pts)
8) 436.419 (10 pts)

ANSWERS TO 'DO YOU UNDERSTAND?'

1) The answer (902.32) is a three-and-two number.
2) The last two digits are 56.
3) The first two digits are 90.
4) Since the markers of the original numbers are 7 and 5, the marker of the answer is 5.

PART FOUR ⟩⟩⟩⟩⟩⟩⟩⟩⟩⟩⟩⟩

13

SUPERFAST
FRACTIONS

Answers to fraction problems should always be presented in their simplest form, i.e., 13 ½ + 5 ⅝ = 19 ⅛, not 18 9/8 or 15 33/8.

Here we enter a new area. Problems involving fractions are not as straightforward as those with integers or decimals, since the two parts of each number really do need to be treated differently. However, with that said, each problem can be broken up into a group of smaller and simpler problems – and the techniques from the integer sections can be used to solve these. The main difficulty here is remembering several numbers at once, although even this should become fairly easy with a litle practice.

14

SUPERFAST FRACTION ADDITIONS

Here are your test problems for the fraction additions section.

Time limit: 6 minutes
1) $15 \frac{2}{5} + 10 \frac{2}{19}$ (4 pts)
2) $13 \frac{5}{13} + 2 \frac{11}{16}$ (4 pts)
3) $11 \frac{4}{9} + 14 \frac{1}{10}$ (4 pts)
4) $1 \frac{4}{7} + 18 \frac{4}{9}$ (4 pts)
5) $19 \frac{8}{11} + 7 \frac{7}{18}$ (4 pts)
6) $2 \frac{1}{3} + 2 \frac{5}{12}$ (4 pts)
7) $7 \frac{1}{3} + 19 \frac{1}{2}$ (4 pts)
8) $16 \frac{5}{8} + 13 \frac{3}{11}$ (4 pts)
9) $2 \frac{7}{10} + 19 \frac{4}{11}$ (4 pts)

It will probably not surprise you to learn that the key to working with fractions is to break down each problem into a small number of integer problems. The main difficulty with fraction additions is the fact that the fraction parts of the numbers rarely use the same units.

A little explanation is now required regarding the terminology of fractions. If a fraction is represented as 'a $\frac{b}{c}$' then 'a' is (not surprisingly) the integer part, 'b' is the numerator and 'c' is called the denominator. For instance, in the number 2 $\frac{8}{9}$, the integer is 2, the numerator is 8 and the denominator is 9.

The first part of the problem is to add the fraction parts of the numbers. This is achieved by multiplying the numerator of each fraction by the denominator of the other and adding the two results together, then multiplying the two denominators together. The new numerator and denominator must now be checked to see if they share any common factors. If they do, then both parts can be divided by the same number (known as 'reducing' or 'simplifying') and thus making them smaller and easier to handle. If the numerator is larger than the denominator, subtract the value of the denominator from the numerator and add an extra 1 to the rest of the addition. As you may have noticed, this is the equivalent of a carry-over in an integer addition.

HERE'S THE PROBLEM: 4 ⁹⁄₁₀ + 6 ⁶⁄₁₉
THIS IS WHAT YOU DO:

The first numerator is 9 and the second denominator is 19, these multiply to give 171. With the second numerator being 6 and the first denominator 10, this gives us 60. 171 + 60 = 231, and so the numerator on the answer is presently set as 231. The two denominators, 10 and 19, multiply to give 190. As this is thus provisionally the denominator for our answer, the numerator (231) is seen to be larger. Consequently, the numerator must be adjusted. 231 – 190 = 41, and so the new numerator is 41. This is a prime number, and the denominator (190) is not divisible by 41, so there is no reduction necessary. The fraction part of the answer is ⁴¹⁄₁₉₀. As the integer parts of the numbers are 4 and 6, they – along with the carry-over of 1 – add to an integer total of 11. The final answer to the problem is thus 11 ⁴¹⁄₁₉₀.

HERE'S THE PROBLEM: 2 ⁵⁄₁₄ + 17 ⁵⁄₈
THIS IS WHAT YOU DO:

Moving through this problem more quickly than the previous one, we see that 5 x 8 = 40,

━━━ CHAMPION'S TOP TIP ━━━

The fraction parts of these additions can add to more than 1, leaving the integer parts needing to be raised by 1. Here's a quick way to check such an addition: multiply the numerator of each fraction by the denominator of the other and add the two results together. Now multiply the two denominators together to make a comparison number. If the comparison is smaller than the sum of the two other multiplications, the two fraction parts do indeed add to more than 1.

14 x 5 = 70, 40 + 70 = 110, and 14 x 8 = 112. Since we now have a numerator which is smaller than the denominator (110 vs. 112), the numerator does not need to be lowered because of its size. However, a quick look reveals that both parts of the fraction are divisible by 2. Reducing both parts by 2, we find the new 'version' of the fraction part of the answer is $^{55}/_{56}$. Adding the integers, i.e., 2 and 17, we find the integer part of the answer is 19. The full answer is thus 19 $^{55}/_{56}$.

One way to check the two parts of the fraction for common factors is as follows: subtract the numerator from the denominator, check to see if the result can be divided by any small numbers, then see if the numerator is divisible by any of these. If the numerator is divisible, the denominator will be divisible also. This method can be extended by finding the difference between the subtraction result and the numerator and checking this result, or even using the results from further such subtractions. Please note that, if further subtractions are used, the denominator will need to be checked along with the numerator.

HERE'S THE PROBLEM: 14 $\frac{4}{15}$ + 17 $\frac{7}{12}$
THIS IS WHAT YOU DO:

Working quickly once again, we have 4 x 12 = 48, 15 x 7 = 105, 48 + 105 = 153 and 15 x 12 = 180. Consequently, the fraction part of the answer is provisionally set as $\frac{153}{180}$. We see that 180 – 153 = 27, which is 3 x 3 x 3. Thus, we need to check the numerator to see if it is divisible by up to three factors of 3. 151/3 = 51 and 51/3 = 17, but 17 cannot be divided by 3. Therefore there are 2 factors of 3, and since 3 x 3 = 9 we must divide each part of the fraction by 9. 153/9 = 17 and 180/9 = 20, and so the fraction part of the answer is $\frac{17}{20}$. Adding the integer parts, i.e., 14 and 17 (with no carry-over from the fraction part), the integer part of the answer is given as 31. Thus the final answer is 31 $\frac{17}{20}$.

HERE'S THE PROBLEM: 5 $\frac{5}{6}$ + 12 $\frac{7}{15}$
THIS IS WHAT YOU DO:

Looking quickly at the fraction part of the sum once again, we see that 5 x 15 = 75, 6 x 7 = 42, 75 + 42 = 117 and 6 x 15 = 90. This gives us, provisionally, the fraction part of the answer as $\frac{117}{90}$. Since 117 is clearly greater than 90, it must be lowered by this amount and a carry-over of 1 applied to the integer part of the addition. 117 – 90 = 27, and so we are left with a fraction of $\frac{27}{90}$.

We could look at 90 -27 = 63, but we can go further this time. Subtracting the numerator from this result, we get 63 – 27 = 36. Another similar subtraction gives us 36 – 27 = 9, and it should be fairly clear that we are not likely to find a lower or simpler number to check. The number 9 is simply 3 x 3, so we now check the numerator for up to two factors of 3. 27 is divisible by 9 (using both factors of 3), as is 90, and so we are able to reduce both parts of the fraction by 9. 27/9 = 3, and 90/9 = 10. The integers add to 17, and the carry-over of 1 raises this to 18. The final answer to this problem is thus 18 $\frac{3}{10}$.

It is possible to check the results from fraction additions as they stand, but it is simpler and easier to check the intermediate results as you go. Since these checks are covered in the sections on working with integers, there is no reason to repeat that part of the work here.

DO YOU UNDERSTAND?

1) What is the fraction part of the answer to $13\frac{1}{3} + 2\frac{3}{8}$?

2) $15\frac{1}{4} + 17\frac{1}{17} = 32\frac{21}{68}$. What is the denominator in the answer?

3) What is the fraction part of the answer to $16\frac{3}{4} + 4\frac{5}{9}$?

4) What is the fraction part of the answer to $2\frac{11}{14} + 2\frac{1}{2}$?

ANSWERS TO THE TIMED TEST PROBLEMS

1) $25 \frac{48}{95}$ (4 pts)

2) $16 \frac{15}{208}$ (4 pts)

3) $25 \frac{49}{90}$ (4 pts)

4) $20 \frac{1}{63}$ (4 pts)

5) $27 \frac{23}{198}$ (4 pts)

6) $4 \frac{3}{4}$ (4 pts)

7) $26 \frac{5}{6}$ (4 pts)

8) $29 \frac{79}{88}$ (4 pts)

9) $22 \frac{7}{110}$ (4 pts)

ANSWERS TO 'DO YOU UNDERSTAND?'

1) $\frac{17}{24}$.

2) 68.

3) $\frac{17}{76}$ (The fractions add to $\frac{93}{76}$, and 93 – 76 = 17).

4) $\frac{2}{7}$ (The fractions add to $\frac{36}{28}$, then 36 – 28 = 8 leaving $\frac{8}{28}$. Reduction by a factor of 4 leaves the fraction as $\frac{2}{7}$).

15

LIGHTNING-FAST FRACTION SUBTRACTIONS

With the fraction additions now 'out of the way' along with two separate groups of subtractions, you have already covered most of the work involved in solving fraction subtractions. This section should cause you no discomfort, so here we go...

As usual, here are some problems to begin with.

Time limit: 7 minutes.

1) $27 \frac{1}{8} - 19 \frac{4}{9}$ (5 pts)
2) $24 \frac{1}{8} - 9 \frac{1}{3}$ (5 pts)
3) $16 \frac{15}{16} - 14 \frac{1}{2}$ (5 pts)
4) $2 \frac{5}{8} - 2 \frac{2}{9}$ (5 pts)
5) $7 \frac{5}{7} - 3 \frac{3}{20}$ (5 pts)
6) $32 \frac{12}{17} - 14 \frac{2}{3}$ (5 pts)

7)   12 ⁶/₁₇ – 2 ¼ (5 pts)
8)   20 ⁴/₇ – 15 ½ (5 pts)
9)   15 ⁵/₇ – 8 ³/₇ (5 pts)

It should be fairly clear that the first step in solving this type of problem is to adjust the fraction parts of the numbers so that they each have the same denominator. Next, we check to see if the second numerator is larger than the first. If it is, then we must add the value of the denominator to the first numerator and reduce the first integer by 1. This is the reverse of the carry-over already seen in the fraction additions section, and is the equivalent of the boost seen in integer subtractions.

HERE'S THE PROBLEM: 22 ⁵/₉ – 10 ½
THIS IS WHAT YOU DO:

- As with fraction additions, we start by multiplying the numerators by each other's denominator. 5 x 2 = 10 and 9 x 1 = 9; since 10 is larger than 9, we do not need to adjust the first numerator. 10 – 9 = 1, and 9 x 2 = 18 (multiplying the denominators); thus the fraction part of the answer is ¹/₁₈.

- Subtracting the integer parts of the numbers, we find 22 – 10 = 12. Consequently, the final answer to this problem is 12 $\frac{1}{18}$.

CHAMPION'S TOP TIP

Some fraction subtractions have a larger fraction part subtracted from a smaller one. Here's how to check if that's what you are facing: multiply the numerator of the first fraction by the denominator of the second, then multiply the numerator of the second by the denominator of the first as a comparison. If this comparison is larger than the result of your first multiplication, the fraction part of the second number is indeed larger than that of the first.

HERE'S THE PROBLEM: 26 $\frac{7}{20}$ – 12 $\frac{11}{13}$
THIS IS WHAT YOU DO:

- Again we start by adjusting the fraction parts of the two numbers. 7 x 13 = 91, and 20 x 11 = 220. As 220 is higher than 91, i.e. the second numerator is greater than the first. We now need to add the value of the denominator to the first numerator. This is 20 x 13 = 260. Since 91 + 260 = 351, we now have a first numerator of 351. The numerator of the answer is thus 351 – 220 = 131, and so the fraction part of the answer is $^{131}\!/_{260}$. Since 131 is a prime number and 260 is not divisible by it, there are no common factors and thus no reduction is necessary.

- The subtraction of the integer parts of the numbers initially gives 26 – 12 = 14, but it must be remembered that the numerator of the first fraction was given a 'boost'; thus the integer part of the answer must be lowered by 1 to give 13. The final answer to the problem is thus 13 $^{131}\!/_{260}$.

HERE'S THE PROBLEM: 6 $\frac{1}{10}$ – 4 $\frac{1}{4}$
THIS IS WHAT YOU DO:

- Working quickly through the first part of the process, we have 1 x 4 = 4 and 10 x 1 = 10. 10 is greater than 4, and so we again add the denominator (10 x 4 = 40) to the first numerator. 40 + 4 = 44, and 44 – 10 = 34, so the fraction part of the answer is presently seen as $^{34}\!/_{40}$.

- It can be seen that both the numerator and the denominator are even, i.e. they can each be divided by 2. But what about other potential factors? If we use the subtraction method given in the fraction additions section, we can find out relatively easily.

- 40 – 34 = 6, and so we already have a small and simple number. The 6 is divisible only by 2 and 3, and so these are the only factors we need to check the numerator for. Since 34 can be divided by 2 but not by 3, then we only need to reduce each of the two parts of the fraction by a factor of 2. Since 34/2 = 17 and 40/2 = 20, the fraction part of the answer is $^{17}/_{20}$.

- The integers appear at first to give a result of 6 – 4 = 2, but again the first numerator received a boost. Consequently, the integer part of the answer must again be lowered by 1. The final answer is thus 1 $^{17}/_{20}$.

- It should come as no surprise to find that the answer to a fraction subtraction can be checked by turning it around and treating it as an addition.

EXAMPLE: 18 $^{6}/_{11}$ – 8 $^{2}/_{15}$ = 10 $^{68}/_{165}$

- Turning this equation around, we have 10 $^{68}/_{165}$ + 8 $^{2}/_{15}$ = 18 $^{6}/_{11}$. Looking at the fraction parts and adjusting them, we get 68 x 15 = 1020,

165 x 2 = 330, 1020 + 330 = 1350 and 165 x 15 = 2475. With an apparent fraction of $^{1350}/_{2475}$, the numerator and denominator – when checked – can each be seen to be divisible by two factors of 3 and two factors of 5. This indicates that each can be divided by 225 (3 x 3 x 5 x 5 = 225), thus reducing the numerator to 6 and the denominator to 11. The fraction part of the answer is thus $^{6}/_{11}$, which we have already seen.

- Did you spot a way to use smaller numbers in this part of the confirmation? The use of smaller numbers generally means easier working and thus a lesser likelihood of making a mistake. The fact is that 165 (the first denominator in the addition) is an exact multiple of 15 (the second denominator). In this case, we need only to multiply the second numerator and denominator by the ratio of the two in order to obtain fractions with identical denominators.

- Here's how the numbers 'pan out' on this one: 165/15 = 11, and so this is the number by which we need to multiply the two parts of the second fraction. 2 x 11 = 22 and 15 x 11 = 165, and so the fraction part of the second number becomes $^{22}/_{165}$. As the denominators of the two fractions are now the same, we can simply add the two numerators. $68 + 22 = 90$, and so the fraction part of the answer is seen as $^{90}/_{165}$. This is much easier to handle than $^{1350}/_{2475}$. The numerator and denominator can each be reduced by a factor of 15 (much easier than 225) to give $^{6}/_{11}$ as before. By adding the integer parts of the numbers, i.e. $10 + 8 = 18$, the final answer is $18 \, ^{6}/_{11}$ – as expected.

- It is true that this simplification was not introduced into the fraction additions section, but it is also the case that fraction additions problems normally don't allow this to be used. Where it *is* usable, the denominators are usually small and simple.

DO YOU UNDERSTAND?

1) What is the fraction part of the answer to $17 \, ^{3}/_{5} – 12 \, ^{7}/_{8}$?

2) What is the denominator in the answer to $9 \, ^{1}/_{6} – 8 \, ^{5}/_{18}$?

3) What is the numerator in the answer to $26 \, ^{19}/_{20} – 7 \, ^{5}/_{7}$?

ANSWERS TO THE TIMED TEST PROBLEMS

1) 7 $\frac{34}{45}$ (5 pts)

2) 15 (5 pts)

3) 2 $\frac{7}{16}$ (5 pts)

4) $\frac{29}{12}$ (5 pts)

5) 4 $\frac{99}{140}$ (5 pts)

6) 18 $\frac{2}{51}$ (5 pts)

7) 10 $\frac{7}{68}$ (5 pts)

8) 5 $\frac{1}{14}$ (5 pts)

9) 7 $\frac{2}{7}$ (5 pts)

ANSWERS TO 'DO YOU UNDERSTAND?'

1) $\frac{37}{45}$

2) 9 (9 $\frac{1}{6}$ – 8 $\frac{5}{18}$ = $\frac{8}{9}$)

3) 33 (26 $\frac{19}{20}$ – 7 $\frac{5}{7}$ = 19 $\frac{33}{140}$)

16

FRACTION MULTIPLICATIONS IN THE FAST LANE!

Here are your fraction multiplications test problems.

As you might already have guessed, these problems are solved by multiplication, addition and a final division. There should be no surprises waiting for you here, since these problems (just like those in the previous two sections) can be solved by the use of techniques which, essentially, have already been covered. It's really not much more than a matter of working out what order to do things in.

Time limit: 15 minutes.

1) 19 $\frac{2}{11}$ x 3 $\frac{1}{6}$ (5 pts)
2) 8 $\frac{16}{17}$ x 7 $\frac{5}{7}$ (5 pts)

3) $9 \frac{2}{3} \times 1 \frac{1}{15}$ (5 pts)
4) $16 \frac{1}{2} \times 3 \frac{2}{5}$ (5 pts)
5) $8 \frac{10}{19} \times 14 \frac{1}{7}$ (5 pts)
6) $19 \frac{5}{8} \times 4 \frac{19}{20}$ (5 pts)
7) $4 \frac{1}{3} \times 9 \frac{9}{10}$ (5 pts)
8) $4 \frac{3}{5} \times 11 \frac{1}{2}$ (5 pts)
9) $13 \frac{3}{14} \times 1 \frac{5}{9}$ (5 pts)

The main method for solving these problems begins by turning each of the two numbers into a simple (if top-heavy) fraction. By multiplying the two numerators to find a new numerator, and then likewise multiplying the denominators, a new top-heavy fraction is created. This can now be treated as an inexact integer division. However, once the integer part of the result has been determined the remainder is *not* used to continue the division; rather it becomes the numerator in the fraction part of the answer. The numerator and denominator should be checked, either now or before the division, for common factors (as seen previously) before the final answer is presented.

HERE'S THE PROBLEM: 4 ¾ X 7 ⅔
THIS IS WHAT YOU DO:

- To present each number as a simple fraction, the 'new' numerator is found by multiplying the integer by the denominator and adding the 'old' numerator; the denominator stays as it is.

- For the first number, 4 x 4 = 16 and 16 + 3 = 19 and so the simple fraction is ¹⁹⁄₄. The second number gives us 7 x 7 = 49 and 49 + 2 = 51, resulting in the fraction ⁵¹⁄₇. Multiplying the numerators, we find 19 x 51 = 969, and the denominators give us 4 x 7 = 28. The answer, expressed as a simple fraction, is thus ⁹⁶⁹⁄₂₈.

- At this point, it is worth checking this fraction to see if there are any common factors. The denominator, 28, is only divisible by factors of 2 (twice) and 7 (once). Since the numerator (969) cannot be divided by either of these, no reduction is possible.

- Dividing 969 by 28 using the method described in the integer divisions section, we see that 96/28 = 3 with a remainder of 12 and that 129/28 = 4 with a remainder of 17. Since this remainder is now to be used as the numerator in the fraction part of the answer, and with the denominator set as 28, the fraction part of the answer is $^{17}\!/_{28}$.

- The integer part has been given as 3 followed by 4, i.e., it is 34. The final answer to this fraction multiplication is therefore 34 $^{17}\!/_{28}$.

- When the numbers are particularly large, this technique can result in the use of large numbers in the multiplication. When the numerators call for a four-by-four multiplication, for instance, things can get rather difficult, especially when you have to keep everything in your head. Remembering a few small numbers can be easier than remembering two very large ones. The secondary method works by turning one multiplication into four. Multiply the two integers, then multiply the integer part of each number by the other's fraction part, and finally multiply the two fraction parts.

HERE'S THE PROBLEM: 37 $^{27}/_{34}$ X 34 $^3/_7$
THIS IS WHAT YOU DO:

- The first step is to multiply the integers. As 37 x 34 = 1258, this is the bulk of the integer answer to the problem.

- Multiplying the first integer by the second numerator, we get 37 x 3 = 111. Using the second denominator, this becomes $^{111}/_7$. Working as before to present this fraction in its simplest terms, we find the number is 15 $^6/_7$. Adding this to the integer already discovered, we find 1258 + 15 = 1273. Our total is now 1273 $^6/_7$.

- Now multiplying the fraction part of the first number by the integer part of the second, we have $^{27}/_{34}$ x 34. As we would be multiplying 27 by 34 and then dividing by the same value of 34, the multiplication and division cancel each other out. This part of the multiplication is therefore exactly 27, and when this is added to the 1273 $^6/_7$ found above we reach a new sub-total of 1300 $^6/_7$.

- As we have already seen, the fraction parts of the numbers are dealt with by multiplying the two numerators and then the two denominators and checking for common factors. This gives a numerator of 27 x 3 = 81 and a denominator of 34 x 7 = 238. The numerator only has factors of 3 (four of them), and the denominator is not divisible by 3. Therefore, there are no common factors, and there is no reduction to be done.

- Using the fraction addition methods already described, we see that 1300 $^6/_7$ + $^{81}/_{238}$ has a second denominator which is an exact multiple of the first. 238/7 = 34, 34 x 6 = 204, and 204 + 81 = 285. The fraction part of the answer now appears to be $^{285}/_{238}$. Since 285 is greater than 238 it must be lowered by this amount, the integer part of the answer being raised by 1 to 1301. As 285 − 238 = 47, and 47 shares no common factors with 238, the fraction part of the answer is $^{47}/_{238}$. The final answer to this problem is thus 1301 $^{47}/_{238}$.

Checking the accuracy of the answer to a fraction multiplication problem is not particularly easy; it is probably better to check your intermediate calculations as you progress through the problem. However, checks can be applied by turning the equation around, returning each of the three terms of the equation to a simple fraction and attempting to divide. This method will be discussed only briefly here as the next section deals with fraction divisions in greater detail.

CHAMPION'S TOP TIP

Here's a way you can check your answer – but it's not easy. Multiply the two denominators in the problem and check to see if the answer has the result as its own denominator. If not, the multiplication result should be a multiple of the answer's denominator. Failing this, the answer is wrong. Multiply the denominator of the answer by the integer and add the numerator, then do this for the second of the numbers in the original problem. Divide the larger number by the smaller one and, if your 'first test' multiplication showed one number was a multiple of the other, multiply your answer by that multiple. Now take this result and look to the first number in the original problem: subtract the numerator, then divide by the denominator; you should now be left with the value of the integer. This will be the confirmation that your answer to the problem is correct.

EXAMPLE: 1 $\frac{1}{13}$ X 8 $\frac{8}{13}$ = 11 $\frac{45}{169}$

Turning this equation around, we get 11 $\frac{45}{169}$ / 8 $\frac{8}{13}$ = 1 $\frac{1}{13}$. Turning each of the three terms into a simple fraction, this becomes $\frac{1904}{169}$ / $\frac{112}{13}$ = $\frac{17}{13}$.

Inverting the divisor (the second number) and multiplying instead of dividing is now the key. The equation we are left with is thus $^{1904}/_{169}$ x $^{13}/_{12}$ = $^{17}/_{3}$. As this is a multiplication, the numerators can be exchanged leaving $^{13}/_{169}$ x $^{1904}/_{12}$ = $^{17}/_{3}$. Since the first two terms can be reduced by factors of 13 and 112 respectively, this now becomes $^{1}/_{3}$ x 17 = $^{17}/_{3}$ which is clearly correct.

DO YOU UNDERSTAND?

1) How does 7 $^5/_6$ x 6 $^5/_7$ appear, in terms of simple fractions, prior to multiplication?

2) Using the second method for multiplication, what are the four numbers you would need to add together to find the answer to 19 $^8/_{11}$ x 17 $^5/_7$?

3) What is the denominator in the answer to 19 $^8/_{11}$ x 2 $^5/_7$?

ANSWERS TO THE TIMED TEST PROBLEMS

1) $60 \frac{49}{66}$ (5 pts)
2) $68 \frac{116}{19}$ (5 pts)
3) $10 \frac{14}{45}$ (5 pts)
4) $56 \frac{1}{10}$ (5 pts)
5) $120 \frac{78}{133}$ (5 pts)
6) $97 \frac{23}{160}$ (5 pts)
7) $42 \frac{9}{10}$ (5 pts)
8) $52 \frac{9}{10}$ (5 pts)
9) $20 \frac{5}{9}$ (5 pts)

The maximum total score so far is 459 points, with another 291 yet to be made available.

ANSWERS TO 'DO YOU UNDERSTAND?'

1) $\frac{47}{6} \times \frac{47}{7}$.
2) The numbers are 323, $16 \frac{2}{7}$, $12 \frac{4}{11}$. and $\frac{48}{77}$.
3) The denominator is 11, not 77. Looking at the first number, 19 x 11 = 209 and 209 + 8 = 217 which is divisible by 7 (the second denominator). The actual answer to the problem is $56 \frac{4}{11}$.

17

SUPER-SPEEDY FRACTION DIVISIONS

Although it may not at first appear to be true, fraction divisions are almost as easy as the multiplications you have just worked through. There is one small trick to make things much simpler; let's see if you can spot it before reading through this section.

Here are your fraction division problems.

Time limit: 15 minutes.

1) $5\,^{13}\!/_{19}$ / $12\,\frac{1}{2}$ (6 pts)
2) $15\,\frac{1}{2}$ / $9\,\frac{1}{4}$ (6 pts)
3) $16\,^{12}\!/_{17}$ / $16\,^{5}\!/_{13}$ (6 pts)
4) $19\,^{9}\!/_{19}$ / $2\,^{2}\!/_{7}$ (6 pts)
5) $18\,^{7}\!/_{8}$ / $10\,\frac{1}{2}$ (6 pts)
6) $5\,^{3}\!/_{10}$ / $1\,^{14}\!/_{17}$ (6 pts)

7) 8 $\frac{5}{17}$ / 10 $\frac{5}{6}$ (6 pts)
8) 13 $\frac{3}{4}$ / 6 $\frac{1}{3}$ (6 pts)
9) 8 $\frac{1}{2}$ / 11 $\frac{3}{11}$ (6 pts)

As I mentioned earlier, the key phrase is usually given as 'invert the divisor and multiply'. To put this another way, turn the divisor (the second number) upside down (by swapping the numerator and the denominator) and then treat the problem as a multiplication instead of a division. The methods given in the fraction multiplications section can then be used to solve the problem.

HERE'S THE PROBLEM: 11 $\frac{2}{7}$ / 12 $\frac{4}{5}$
THIS IS WHAT YOU DO:

Turning each term into a simple fraction, this problem becomes $\frac{79}{7}$ / $\frac{64}{5}$. Inverting the divisor and turning the problem into a multiplication leaves the problem appearing as $\frac{79}{7}$ x $\frac{5}{64}$. The next stage is, of course, to multiply the two numerators and then the two denominators. Since 79 x 5 = 395 and 7 x 64 = 448, the result appears to be $\frac{395}{448}$. Upon checking for common factors, it can be seen that none exists; thus $\frac{395}{448}$ is the final answer to the problem.

HERE'S THE PROBLEM: 19 $\frac{1}{9}$ / 16 $\frac{5}{6}$
THIS IS WHAT YOU DO:

With this problem, the first number becomes $\frac{172}{9}$ and the second is $\frac{101}{6}$. Upon inverting the divisor and thus turning the division into a multiplication, the problem then becomes $\frac{172}{9}$ x $\frac{6}{101}$. Exchanging the numerators gives us $\frac{6}{9}$ x $\frac{172}{101}$. The two parts of the fraction $\frac{6}{9}$ can each be reduced by a factor of 3, giving $\frac{2}{3}$ and thus leaving the multiplication as $\frac{2}{3}$ x $\frac{172}{101}$. The result of this multiplication is thus $\frac{344}{303}$. Treating this as an inexact integer division, we see an answer of 1 with a remainder of 344 − 303 = 41. As 41 and 303 share no common factors, no reduction is possible. The final answer to the problem is therefore 1 $\frac{41}{303}$.

HERE'S THE PROBLEM: 10 $\frac{1}{17}$ / 3 $\frac{2}{3}$
THIS IS WHAT YOU DO:

- Presenting this problem in terms of simple fractions gives us $\frac{171}{17}$ / $\frac{11}{3}$. Inverting the divisor then leaves the problem appearing as $\frac{171}{17}$ x $\frac{3}{11}$. Multiplying the two numerators and the two denominators gives us $\frac{513}{187}$ as a provisional answer.

- Applying the inexact integer division technique, we reach an answer of 2 with a remainder of 139. Since 139 and 187 do not share any common factors, there is no reduction to be performed. Therefore the fraction part of the answer is $^{139}/_{187}$, which means the final answer to the problem is $2\ ^{139}/_{187}$.

- Answers to fraction division problems, just like answers to fraction multiplications, are not easily checked for accuracy. Again it is probably best to check the results of your intermediate calculations as you progress through the problem, but you can also check your answer if you wish. It will probably come as no surprise to learn that the method of checking the answer to a division is to turn it around and treat it as a multiplication.

EXAMPLE: 10 $^5/_6$ / 14½ = $^{65}/_{87}$

Turning this equation around leaves it appearing as $^{65}/_{87}$ x 14 ½ = 10 $^5/_6$. Showing each number as a simple fraction, this becomes $^{65}/_{87}$ x $^{29}/_2$ = $^{65}/_6$. If we exchange the numerators again we get $^{29}/_{87}$ x $^{65}/_2$ = $^{65}/_6$. Since 29 and 87 can each be divided by 29, the two parts of the first fraction can each be reduced by a factor of 29. This now leaves $\frac{1}{3} \times \frac{65}{2}$ = $\frac{65}{6}$. Since 1 x 65 = 65 and 3 x 2 = 6, this is clearly correct and thus the answer ($\frac{65}{6}$) to the original problem (10 $\frac{5}{6}$ / 14 $\frac{1}{2}$) is correct.

— CHAMPION'S TOP TIP —

Since a division can be checked by turning it around and treating it as a multiplication (we have already seen this with integers and decimals), try this: once you have found an answer to a fraction division, turn the problem around and treat your answer as the first number in a multiplication. With the first number in the original division problem as the answer to the division, you can use Champion's top tip on fraction multiplications to check this answer. If it works, then your answer to the original division problem is correct.

DO YOU UNDERSTAND?

1) Faced with the problem 5 $\frac{1}{3}$ / 11 $\frac{9}{13}$, how would the problem appear as a multiplication of two simple fractions?

2) In a similar manner, how would the problem 18 $\frac{4}{5}$ / 9 $\frac{1}{2}$ appear?

3) What multiplication, again shown in terms of simple fractions, would you use to check the answer to 17 $\frac{3}{4}$ / 3 $\frac{5}{6}$ = 4 $\frac{29}{46}$?

ANSWERS TO THE TIMED TEST PROBLEMS

1) $^{216}/_{475}$ (6 pts)
2) $1\ ^{25}/_{37}$ (6 pts)
3) $1\ ^{1}/_{51}$ (6 pts)
4) $8\ ^{79}/_{152}$ (6 pts)
5) $1\ ^{544}/_{639}$ (6 pts)
6) $2\ ^{281}/_{310}$ (6 pts)
7) $^{846}/_{105}$ (6 pts)
8) $2\ ^{13}/_{76}$ (6 pts)
9) $^{187}/_{248}$ (6 pts)

ANSWERS TO 'DO YOU UNDERSTAND?'

1) $^{16}/_{3}$ x $^{13}/_{152}$.

   This may also be exchanged to $^{13}/_{3}$ x $^{16}/_{152}$ and reduced to $^{13}/_{3}$ x $^{2}/_{19}$.

2) $^{94}/_{5}$ x $^{2}/_{19}$.

3) $^{213}/_{46}$ x $^{23}/_{6}$ = $^{71}/_{4}$.

PART FIVE ⟫⟫⟫⟫⟫⟫⟫⟫⟫

18

QUICK CALCULATIONS: RAPID REMAINDERS

Remainders are, as the name suggests, the left-over numbers after inexact integer divisions have been performed. For example, 461 / 52 = 8 with 45 left over, so the answer to 461 / 52 as a remainder problem is 45. Each of these problems is very similar to the final stage of a fraction multiplication problem.

Here are some remainder problems for you to try.

Time limit: 19 minutes.

1) 570 / 29 (2 pts)
2) 26013 / 264 (4 pts)
3) 189830 / 987 (5 pts)
4) 489371 / 1756 (6 pts)
5) 838108 / 5539 (7 pts)

6) 5462898 / 3180 (8 pts)
7) 160620600 / 8054 (9 pts)
8) 474636692 / 39715 (10 pts)

You should by now have realised that we have already encountered a few low-value remainder problems in earlier sections. Now it is time to look at them in more detail and using higher numbers.

Remainder problems can all be treated in the manner described earlier, but it is advisable to keep a running total of how many times you subtract the divisor; this is for checking purposes.

HERE'S THE PROBLEM: 148669 / 956
THIS IS WHAT YOU DO:

- The first step is to find how many digits from the original number (in this case 148669) are required for the first part of the division. The divisor (956) has three digits, and the first three digits of 148669 (i.e., 148) do not provide a large enough number; therefore four are required.

- The first part of the division is thus 1486 / 956. It is easy to see that 956 can only be subtracted from 1486 once, leaving 1486 – 956 = 530. This means our sub-total of subtractions is now 1, and the digits '530' must now be followed with a further digit.

- Since we have already used the first four digits from the original number 148669, the fifth digit must now be introduced. As this digit (a 6) has to follow the remainder from the first part of the division, i.e., 530, this means the next part of the division is given as 5306 / 956. Using the technique already described, you should soon see that 956 can be subtracted five times from 5306 (as 5 x 956 = 4780) leaving a remainder of 5306 – 4780 = 526. As the divisor has now been subtracted a further five times, the sub-total is now 1 + 5 = 6.

- Following the same process as used in the previous step, we find that the next part of the problem is given by the division of 5269 / 956. This is in fact the last division required in solving this particular problem as it involves the final digit of the large number. It should be clear that the 956 can again be subtracted five times (5 x 956 = 4780), this time leaving a remainder of 5269 – 4780 = 489. As this was the final division within this problem, the remainder of 489 is the final answer. With another five sub-tractions of 956, our final total is 6 + 5 = 11.

- The method used to check the accuracy of the answer to a remainder problem such as this is based on markers. Your answer should be correct if you have worked carefully and checked your calculations as you have worked your way through the problem, but mistakes can always be made – it's best to check again, if you can.

- Multiply the marker of the divisor by the marker of the total number of subtractions you have made, then add the marker of your answer. The result should have the same marker as the original number.

EXAMPLE: 148669 / 956 HAS A REMAINDER OF 489 AFTER 11 SUBTRACTIONS

This, as you can see, is the problem we have just solved – and now it's time to check the answer. The divisor is 956, with a marker of 2, and the total of subtractions is 11 which also has a marker of 2. The answer (489) has a marker of 3, and so we have 2 x 2 = 4 and 4 + 3 = 7. We should thus find that the number 148669 has a marker of 7. Since 1

+ 4 + 8 + 6 + 6 + 9 = 34 and 3 + 4 = 7, this is shown to be the case. Consequently, the answer passes the test – and we thus conclude that our answer of 489 appears to be correct, which indeed it is.

Let's go through that once more, but with larger numbers.

HERE'S THE PROBLEM: 1393556 / 8964
THIS IS WHAT YOU DO:

- The first step, as before, is to find how many digits are required in order to be able to cope with the divisor, which is now 8964 (four digits long). The first four digits of 1393556 are not enough, and so we use five. This gives 13935 / 8964. Since 8964 x 2 = 17928, which is too high, we can see that 8964 can only be subtracted once to leave a remainder of 13935 – 8964 = 4971. The sub-total of the subtractions now stands at 1.

- Having used the first five digits of 1393556, we bring down the next digit (the sixth, a 5) to follow the remainder (4971) to give the number 49715. This leaves us with the division 49715 / 8964, which is quite a knotty little problem in its own right.

- Looking at this part of the problem, we can see that 5 x 8964 = 44820 (which is OK) and 6 x 8964 = 53784 (too high). So we can subtract 8964 five times from 49715 to leave a remainder of 49715 – 44820 = 4895. The subtractions sub-total moves on to 1 + 5 = 6. Introducing the next digit (the seventh, a 6) to follow the remainder from above, we now have the division 48956 / 8964. This is the last division as it involves the last digit from the original number of 1393556. Again we find that 8964 can be subtracted five times (5 x 8964 = 44820) leaving a remainder, and thus the final answer, of 4136. The final total of subtractions is 6 + 5 = 11.

- Now to check the answer: the marker of the subtractions total (11) is 2, that of the divisor (8964) is 9, and that of the answer (4136) is 5. As 2 x 9 = 18 and 18 + 5 = 23 which has a marker of 5, we should expect the original number (1393556) to have a marker of 5 also. It has, and so the answer passes the test.

CHAMPION'S TOP TIP

Since these problems are essentially divisions, there are implications from what was said about multiples of 3 and 9 in the integer divisions section. If the divisor (the smaller number) is divisible by 3, then counts of 1, 4 and 7 will show the same results (as would counts of 2, 5 and 8 or 3, 6 and 9). However, a divisor which is divisible by 9 provides an advantage; you won't need to count the subtractions at all, since the marker of the answer must always be the same as that of the original number.

DO YOU UNDERSTAND?

Which of the following problems has been supplied with the wrong answer?

1) 774617017/80392 leaves 40097 after 23 subtractions.

2) 313709830/4617 leaves 3248 after 32 subtractions.

3) 2614149/9950 leaves 7249 after 10 subtractions.

ANSWERS TO THE TIMED TEST PROBLEMS

1) 19 (2 pts)
2) 141 (4 pts)
3) 326 (5 pts)
4) 1203 (6 pts)
5) 1719 (7 pts)
6) 2838 (8 pts)
7) 7732 (9 pts)
8) 2727 (10 pts)

ANSWER TO 'DO YOU UNDERSTAND?'

Remainder problem 2) has been supplied with the wrong answer. Here, the markers of the subtractions total, the divisor and the given answer are 5, 9 and 8 respectively. Since 5 x 9 = 45 and 45 + 8 = 53 with a marker of 8, we should expect the original number (in this case 313709830) to have a marker of 8 also. As 3 + 1 + 3 + 7 + 0 + 9 + 8 + 3 + 0 =34 and 3 + 4 = 7, we see that the marker of the original number is actually 7, which disagrees with the check – and so the answer as given (3248) is wrong. The correct answer to this problem is 3148.

19

THE FAST WAY TO HANDLE PRIME FACTORS

A brief explanation: the prime factors of a number are the prime numbers which multiply together to give the original number. A prime number is one which cannot be divided by any number other than itself and 1. If a prime factor is used more than once, it should be listed the relevant number of times.

For example, 420 = 2 x 2 x 3 x 5 x 7, and so the answer for 420 is 2, 2, 3, 5, 7.

Finding the list of prime factors of a number is practically the same as a series of integer divisions. However, there are two 'extra' problems: firstly you need to perform each division in turn in order to find what number you are starting with on the next stage; secondly you must work out for yourself what number you are dividing by.

Find the prime factors for the following numbers.

Time limit: 12 minutes.

1) 92 (2 pts)
2) 782 (4 pts)
3) 4692 (5 pts)
4) 8364 (6 pts)
5) 22724 (7 pts)
6) 50286 (8 pts)
7) 185150 (9 pts)
8) 5143996 (10 pts)

You should by now be able to find the prime factors of quite fair-sized numbers without any difficulties, since the list of prime factors is produced by a series of integer divisions.

HERE'S THE PROBLEM: FIND THE PRIME FACTORS OF 11730
THIS IS WHAT YOU DO:

- 11730 / 2 = 5865, which is not divisible by 2.

- 5865 / 3 = 1955 which is not divisible by 3.
- 1955 / 5 = 391 which is not divisible by 5, 7, 11 or 13.
- 391 / 17 = 23 which is a prime number.
- The list of prime factors is thus 2, 3, 5, 17, 23.
- As the numbers get larger, the capacity for mistakes increases dramatically. It can therefore be a considerable advantage to be able to find how many factors of 2 are held within the number. This is done by finding out whether the number is divisible by 2, 4, 8, 16, etc. This not as tricky as it might seem. If you wish to find if the number is divisible by (for example) 8, i.e., if it has three factors of 2, you only need to check the last three digits in the number. Likewise, to find if the number can be divided by 32 (five factors of 2), you need to check the last five digits. Of course, you can still check the whole number if it does not hold the required number of digits.
- If the check is successful, there is a simplified method for checking 'the next level' (i.e.; 16 after 8 or 32 after 16). Having divided the 'limited' number by the relevant power of 2, compare the result with the

preceding digit of the main number. If they are either both odd or both even, the number does indeed have *at least* one further factor of 2. If either is odd and the other is even, then there are no more factors of 2.

HERE'S THE PROBLEM: FIND THE PRIME FACTORS OF 71944
THIS IS WHAT YOU DO:

- It is obvious that the number is even, i.e. it has at least one factor of 2, but how many factors of 2 does it have? The last two figures (44) give a number which is clearly divisible by 4 (two factors of 2), but what about dividing by 8 (three factors of 2)?

- Since we are now looking at three factors of 2, we need to look at the last three digits of 71944. Dividing 944 by 8 we find exactly 118, i.e., the division works, and so the whole number (71944) is divisible by 8. The result is even, but the preceding digit (1) is odd – a mismatch. Therefore, we see that, although 71944 *has* three factors of 2, it does *not* have a fourth.

- Bearing this in mind, we can list three factors of 2 and then set about finding the prime factors of 8993 (as 71944 / 8 = 8993). These remaining factors are 17, 23 and 23 thus leaving a list of 2, 2, 2, 17, 23, 23 as the answer.

- Once the factors of 2 have been determined, it's time to find how many (if any) factors of 3 there are. This is also relatively simple. First, find the marker of the number. If the marker cannot be divided by 3, then neither can the original number. If the marker is 3 or 6, then there is one – and only one – factor of 3 present. However, if the marker is 9 then there are *at least* two factors of three and possibly more; divide the number by 9 and then check the result for further factors of 3.

CHAMPION'S TOP TIP

When you have found the number of factors of 2 and 3 in the number, you can record these as the first part of your answer & set about finding higher factors to divide by. When you have such a larger factor, try dividing by this number first – and then by your factors of 2 and 3. This can be easier, since you will usually end up needing to remember a smaller number in your head – thus giving you a better chance to find the right answer.

HERE'S THE PROBLEM: FIND THE PRIME FACTORS OF 123786
THIS IS WHAT YOU DO:

- Seeking factors of 2, we find that there is only one. 123786 / 2 = 61893, and so we check 61893 for factors of 3. The marker of this number is 9, so there are at least two factors of 3. Taking note of these, we divide 61893 by 9 and get 6877. The marker of this 'new' number is 1, which is clearly not divisible by 3 – and so there are no more factors of 3 to consider. The remaining factors are in fact 13, 23 and 23, which means the list presented as the final answer is 2, 3, 3, 13, 23, 23.

- Factors of 5 can be sought using a similar method to that used for seeking factors of 2. For example, you only need to check the last two digits if you are checking whether or not the number is divisible by 25 (two factors of 5). Likewise, if there are three factors of 5 present then the last three digits must be divisible by 125 (i.e., 125, 250, 375, 500, 625, 750, 875 or 000).

HERE'S THE PROBLEM: FIND THE PRIME FACTORS OF 203550

THIS IS WHAT YOU DO:

- Having found and eliminated one factor each of 2 and 3, we are left looking for the prime factors of 33925. Since the last two digits give a number (25) which is clearly divisible by 25, we see that there are at least two factors of 5 in the number. As 925 is not divisible by 125 (three factors of 5) then there is *not* a third factor of five in the number. 33925 / 25 = 1357, and so this is the number which we investigate to find the remaining prime factors. These are 23 and 59, giving a final list of 2, 3, 5, 5, 23, 59.

- Notice how we have been checking for prime factors in ascending order (2, then 3, then 5). The next three prime numbers are 7, 11 and 13. These three potential prime factors can each be detected by trying to divide the main number by the relevant prime number, but supposing the main number still has four or more digits? Is there a way to make the divisibility check easier? Well, you've probably already guessed that I would not have bothered to ask the question unless the answer was 'yes'.

- The trick is based on the fact that 7 x 11 x 13 = 1001. Because of this fact, long numbers can be broken up (from the right-hand end) into blocks of three digits and the difference (or differences) compared with the prime factor concerned. This method will not reveal if any of the potential prime factors is involved more than once, but the same method can be applied to the result of a division. It can't reduce the number of divisions you need to make, but it can take some of the pressure off finding out whether or not a division will actually work.

HERE'S THE PROBLEM: FIND THE PRIME FACTORS OF 275823912
THIS IS WHAT YOU DO:

- This number bears three factors of 2 and a single factor of 3. These factors multiply to give a total of 24, and 275823912 / 24 = 11492663. Breaking off a block of three digits from the right-hand end, we get the numbers 11492 and 663. As 11492 – 663 = 10829, this number (which could have been obtained by subtracting 1001 x 663 from 11492663 and then dividing by 1000) can be investigated. Since 10829 is still longer than three digits, the same process can be applied again.

- Breaking off another three digits from the right hand end we get the numbers 10 and 829. As 829 – 10 = 819, we now investigate the number 819 (equal to 10829 – 10010); much easier than dealing with 11492663. 819 is divisible both by 7 and by 13, but not by 11. We can therefore see that the number 11492663 has at least one factor each of 7 and 13 but no factors of 11.

- As 11492663 / 7 = 1641809, we now check this new number for further factors of 7. Breaking off the 'first' block of three digits, we find 1641 – 809 = 832 which is not divisible by 7; therefore there are no further factors of 7 to consider. We already know that there are no factors of 11 and at least one factor of 13, and so we now divide 1641809 by 13 giving us 126293.

- Applying the same process to this number, we find the numbers 126 and 293. Since 293 – 126 = 167, which is not divisible by 13, there are no further factors of 13 to be considered.

- The next prime number to try is 17, by which 126293 is indeed divisible. As 126293 / 17 = 7429, we now check to see if this number is also divisible by 17. It is: 7429 / 17 = 437. This is 19 x 23, and so the final list of prime factors of 275823912 is seen as 2, 2, 2, 3, 7, 13, 17, 17, 19, 23.

- Towards the end of the calculations, especially when relatively high prime factors are involved, it can become clear when there are no more than two prime factors to find. In the last sample problem, for instance, division by the second prime factor of 17 leaves a result of 437. Since there can be no remaining prime factors lower than 17, and 17 x 17 x 17 is obviously much greater than 437 (it is actually 4913) then there cannot be as many as three prime factors left to find.

- In order to pin down the final two factors, we can use a technique known as 'the difference of two squares'. This is a method which is normally used for multiplying two small numbers, but it is quite effective here.

- The difference of two squares works like this:

two numbers can be multiplied together by finding the average, calculating the square (multiplying the number by itself), subtracting the lower number from the average and finding the square of the result, and finally subtracting the smaller square from the larger one. For example, multiplying 23 x 29, the average is 26 (half-way between the two), and the average minus the lower number is 26 – 23 = 3. Since 26 x 26 = 676, 3 x 3 = 9 and 676 – 9 = 667, this indicates that the answer is 667. This is the correct answer.

- This method can be turned around in the following manner: take the number you are investigating and add (individually, not successively) squared numbers of increasing sizes (1, 4, 9, 16, 25, etc.) and see if you can find another perfect square as a result.

HERE'S THE PROBLEM: FIND THE PRIME FACTORS OF 948532
THIS IS WHAT YOU DO:

- Using the methods already described, we can find two factors of 2 and one of 13, leaving the number 18241. This cannot be divided by 13, but it is divisible by 17. As 18241 / 17 = 1073, and 17 x 17 x 17 is clearly much greater than 1073 (it is 4913, as seen before) we can see that there can be no more than two prime factors left to identify.

- Now we start adding small squares to this number to try to find another square. 1073 + 1 = 1074 (no good), 1073 + 4 = 1077 (likewise), and 1073 + 9 = 1082 (another failure). However, 1073 + 16 = 1089, and 1089 = 33 x 33. As 16 = 4 x 4, this indicates that the average of the two remaining factors is 33 and the difference between the lower number and the average is 4. Consequently, the numbers in question are (33 – 4 = 29) and (33 + 4 = 37). To check this, try multiplying 29 x 37 and you will see that the answer is 1073.

Please note that the number under investigation might already be a perfect square, in which case the square root will provide each of the two remaining factors. If the point is reached where the lower of the two attempted factors would be lower than the last prime factor to be identified, then the number you are now investigating is a prime number (unless, of course, you have missed something earlier). If you think the final number looks unusually large, check it against some of the lower prime numbers. You may find an error and be able to correct it, or your calculations may show that you got it right in the first place and the number really *is* that big.

The method for checking the accuracy of the answer to a prime factors question is quite predictable; multiply all the prime factors together to see if you get back to the original number.

EXAMPLE: THE PRIME FACTORS OF 43263 ARE 3, 3, 11, 19 AND 23

It is generally best to start with the largest numbers and work back along the line; otherwise you will end up by multiplying a large number by a large number (in this instance, 1881 x 23). It's much easier to multiply 23 by 19 than it is to multiply it by 1881. 23 x 19 = 437, 437 x 11 = 4807, and we can use both factors of 3 together and multiply by 9 (since 3 x 3 = 9). As 4807 x 9 = 43263, the original number, the list given as the answer appears to be accurate. A final glance at the list shows that they are indeed all prime numbers and so the answer given is correct.

DO YOU UNDERSTAND?

1) How many factors of 2 are there in 55016?
2) How many factors of 3 (if any) are there in 203550?
3) The number 20748 has yielded prime factors of 2, 2, 3 and 7 leaving the number 247. What are the remaining prime factors?

ANSWERS TO THE TIMED TEST PROBLEMS

1) 2, 2, 23 (2 pts)
2) 2, 17, 23 (4 pts)
3) 2, 2, 3, 17, 23 (5 pts)
4) 2, 2, 3, 17, 41 (6 pts)
5) 2, 2, 13, 19, 23 (7 pts)
6) 2, 3, 17, 17, 29 (8 pts)
7) 2, 5, 5, 7, 23, 23 (9 pts)
8) 2, 2, 11, 13, 17, 23, 23 (10 pts)

ANSWERS TO 'DO YOU UNDERSTAND?'

1) There are 3 factors of 2 since 016 / 8 = 2
   which is even, whereas the preceding digit
   (5) is odd.

2) There is one single factor of 3, since the
   marker of the original number is 6.

3) The last two factors are 13 and 19. These
   numbers are found by using the difference
   of two squares in the following way: 247 + 9
   (from 9 = 3 x 3) = 256, and 256 = 16 x 16.
   The average of the two numbers is thus 16
   showing a difference of 3 between
   the average and each factor, so the factors are
   (16 – 3 = 13) and (16 + 3 = 19).

SQUARE ROOTS
THE RAPID WAY

As you should know by now, the square root of a number is the number which must be multiplied by itself to give the original number. As an example, if you multiply 17 x 17 the answer is 289, and so the square root of 289 is 17.

Find the square roots of the following numbers; each problem has an exact answer.

Time limit: 11 minutes.

1) 9409 (2 pts)
2) 69696 (3 pts)
3) 1550025 (5 pts)
4) 19989841 (6 pts)
5) 213598225 (7 pts)
6) 3143732761 (8 pts)

7) 13960131409 (9 pts)

8) 91658168001 (10 pts)

To tackle a square root problem, start by breaking down the original number into pairs of digits *from the right-hand end*. The number of pairs (including the single digit, if there is one) will show how many digits there are in the answer. The first two pairs will lead you to the first two digits of the answer, and will also help you to identify the next digit. If you wish, you can start by looking only at the first pair.

HERE'S THE PROBLEM: GIVE THE SQUARE ROOT OF 69169
THIS IS WHAT YOU DO:

- Breaking down this number as described leaves 6 91 69. The first two pairs (actually a single and a pair) give the number 691. Since this is between 676 (26 x 26) and 729 (27 x 27), the first two digits must be 26. As there are three pairs in total, there are three digits in the answer – meaning there is only one left to find.

- As the last digit of the square is a 9, the last

digit of the root must be either a 3 or a 7 (this can be seen from 3 x 3 = 9 and 7 x 7 = 49; compare with the checking used in integer multiplications). Since 691 is much closer to 676 than it is to 729, the answer must therefore be closer to 260 than to 270. As we have already seen that the final digit has to be either a 3 or a 7, this digit must be a 3 giving the final answer as 263.

- To deal with larger numbers, the 'double and divide' method (my name for it, nothing official) is that which is most commonly used. Here's how it works. First, find the first one or two digits as above – I shall assume the search is for two digits rather than one. Subtract the square of this two-digit number from the number formed by the first two pairs, then follow the result with the next pair of digits. This gives you the next number to work with.

- To find the number to divide this by, you should *double* the number formed by the first two digits and add on one more digit. Here's the tricky bit: working out which digit to use. If you regard this digit as q, and the doubled number is ab, then the value of q should be the highest that allows the subtraction of q x abq. Continuing in the same manner (doubling the result so far and adding another digit), further digits can be identified by this method until the last pair is used – and so the last digit of the answer is found.

HERE'S THE PROBLEM: GIVE THE SQUARE ROOT OF 23107249
THIS IS WHAT YOU DO:

- Breaking this number into pairs of digits, we see 23 10 72 49. This tells us that there are four digits in the answer. Taking the first two pairs, we find the number 2310. This is between 2304 (48 x 48) and 2401 (49 x 49), and so the first two digits of the answer are 48.

- Since 2310 – 2304 = 6, and the next pair of digits gives us 72, the next number to work with is 672. Now to find out what to try to divide it by: the only digits we have so far in the answer are the first two, i.e., 48, and 2 x 48 = 96. Therefore we need to find the highest value of q which will allow the subtraction of q x 96q from 672. This clearly can only be 0 since even a 1 would be too high (1 x 961 = 961 which is higher than 672). Since any number multiplied by 0 is 0 and any number minus 0 has itself as a remainder, the remainder at this stage is 672.

- So far we have found the first two digits to be 48 and we have added the digit 0 to give 480 at this point. The remainder from the step shown above is 672, and the next (and last) pair from the original number is 49, so we must now work with the number 67249. To find out what we should try to divide this number by, we do the same as before. Doubling our partial answer of 480 gives us 960, and so we must find the highest value of q which allows the subtraction of q x 960q from 67249. The methods shown in the integer divisions section help us to see that this value of q appears to be about 7. Testing this, we find 9607 x 7 = 67249 and so the 7 is an exact fit. The final answer to the square root of 23107249 is thus 4807. Let's see that method in action again.

HERE'S THE PROBLEM: GIVE THE SQUARE ROOT OF 38217124
THIS IS WHAT YOU DO:

- This number breaks into pairs to give us 38 21 71 24. This tells us that there are four digits in the answer. The first two pairs give us 3821, which is between 3721 (61 x 61) and 3844 (62 x 62) and so the first two digits of the answer are 61.

- Since 3821 – 3721 = 100 (the remainder) and the next pair from the original number gives us 71, the next number to work with is 10071. Now we need to find out what to try to divide this number by. The partial answer so far is 61, and 61 x 2 = 122; we need to find the highest value of q which will allow subtraction of q x 122q from 10071. Taking one digit off each of the long numbers to allow an approximation we find that 1007 / 122 is between 8 and 9. Trying 9 as the value of q, we see that 9 x 1229 = 11061, which is too high. However, 8 x 1228 = 9824, which is permissible. So we have found the third digit to be 8, giving a partial answer so far of 618, and the remainder from this stage is 10071 – 9824 = 247.

- We now have one more pair left from the original number, so we have one digit left to find in the answer. Following the remainder (247) with the final pair (24) leaves the number 24724. Doubling the partial answer (618) gives 1236, and so we need a value of q which will allow subtraction of q x 1236q from 24724. Again shedding digits to find an approximation, the value of q = 2 practically leaps off the page. Testing this, we see 2 x 12362 = 24724 which is another exact fit. The final digit in the answer is 2, and thus the final answer is 6182.

- There are methods, both rough and ready as well as somewhat more refined, of estimating the value of a square root upon finding a partial result. There is also a system for finding the last two digits of an exact square root very quickly. I'll now show you all three techniques.

LINEAR INTERPOLATION

When you have reached a partial answer to a square root problem, you can use what is known as linear interpolation to estimate the final result. When used in line with the 'rule of quadrants', which will be described later, this can give you an accurate answer to most square root problems with original numbers of up to eight digits in length, even if you only find the partial answer from just the first two pairs.

Once you have your partial answer and the associated remainder, you can compare the size of the remainder with the size of the gap between the square of your partial answer and the next square up. This gap will always be one greater than double your partial answer.

HERE'S THE PROBLEM: GIVE THE SQUARE ROOT OF 1071842121
THIS IS WHAT YOU DO:

- This number breaks down into 10 71 84 21 21, from which the first two pairs give the number 1071. Since this is between 1024 (32 x 32) and 1089 (33 x 33), the first partial answer is found to be 32. The remainder is 1071 – 1024 = 47, and this time we also look at the gap as mentioned previously. Since the partial answer is 32, the gap is 32 x 2 + 1 = 65.
- The remainder is 47 and the gap is 65, and

so we divide 47 / 65 = 0.723 (approx.). As there are five digits in the answer and we have so far found only two, there are three more to be found. We thus scale up the approximation of 0.723 by moving the decimal point these three places to the right, leaving us with 723. Placing this after the partial answer of 32, we reach an estimated answer of 32723.

- By going one step further before we use this method of estimation, we can make things much more accurate. Using the double-and-divide as before, we reach a partial answer of 327, a remainder of 255 and two digits left to find. The partial answer of 327 gives us a gap of 327 x 2 + 1 = 655, and the division 255 / 655 = 0.39 to the required two decimal places. This leaves 39 to be added to the end of our partial answer of 327, giving a new estimated answer of 32739 which is actually correct.

REDUCTION FROM FIRST ESTIMATE

Once you have a rough idea of the square root (from the partial answer), the original number can be divided by this estimate and the average taken of this result and the initial estimate.

HERE'S THE PROBLEM: GIVE THE SQUARE ROOT OF 78287104
THIS IS WHAT YOU DO:

- Moving a little more quickly now, we reach a partial answer of 88, a remainder of 84, a gap of 177 and two digits to be found. In view of the remainder and the gap, the estimated root will be slightly below 8850. Dividing 7828704 by 8850 gives almost exactly 8846, and the average of 8850 and 8846 is 8848. This is therefore the estimated answer, and it is indeed correct.

- You can apply a further refinement to this method especially when the answer to the square root problem is not an exact number. Take the average as before, but this time compare the new estimate with the first. Find the ratio of the difference to the lower number, halve this ratio, then reduce the new estimate by this fraction of the difference between the estimates.

HERE'S THE PROBLEM: GIVE THE SQUARE ROOT OF 1288 TO 2 DECIMAL PLACES
THIS IS WHAT YOU DO:

The square root of this number is close to 36, so we take 36 as the first estimate. Dividing 1288 by 36 gives 35.7777, etc. and thus the average (and the new estimate) is 35.8889 to four decimal places. Since 36 − 35.8889 = 0.1111, this being $\frac{1}{323}$ of the lower of the two estimates, and with half of $\frac{1}{323}$ being $\frac{1}{646}$, we 'lose' $\frac{1}{646}$ of 0.1111 from our present estimate. To four decimal places, this leaves a final estimate of 35.8887. Reducing this to two decimal places, this gives a final answer of 35.89 which is correct. To show how accurate this system of estimation can be, it is worth noting that the square root of 1288 to 4 decimal places is indeed 35.8887.

RULE OF QUADRANTS

- This system can be used to find the final two digits of an *exact* square root. Once you have a partial answer to an exact square root with just two digits left to find, you can use the remainder to see which 'quadrant' (01–25, 25–50, 50–75 or 75–99) the final two digits will lie in. If the remainder is less than half the partial answer, the digits will be in the first quadrant. Between half and the whole partial answer, they will be from the second quadrant. A remainder of between one and one-and-a-half times the partial answer will see the final digits coming from the third quadrant, and the fourth quadrant is indicated by a remainder of more than one-and-a-half times the partial answer.

- If the original number ends with a number of zeros, these can be disregarded temporarily while finding the root, then adding half the number of zeros on to the end of the rest of the answer. If the number ends with a 5, then so does the root – and the methods of estimation given above will lead you to the correct answer. However, the last two non-zero digits – if they *don't* end in a five – can only be produced by one number in each quadrant. You don't need to know the squares of all the numbers from 1 - 100, but knowing those from 1 - 25 will help. However, these squares are not difficult to calculate if you don't know them.

- Once you have identified both the quadrant

the final two digits are from and the number in the first quadrant which can create the last two digits of the original number, the rest is simple. Using n to represent the appropriate number from the first quadrant: if the final two digits come from the first quadrant, the two-digit number to finish with is n. A second quadrant number would be 50-n, a third quadrant number would be 50+n and a fourth quadrant number would be 100-n.

HERE'S THE PROBLEM: GIVE THE SQUARE ROOT OF 6342051769
THIS IS WHAT YOU DO:

Using methods already discussed, we find a three-digit partial answer of 796, a remainder of 589 and two digits left to find. The last two digits of the original number are 69, and the only number in the 1–25 range (the first quadrant) which can be responsible is 13 (since 13 x 13 = 169). The value of n is thus 13. The remainder (589) is less than the partial answer (796) but more than half its value, indicating the answer lies in the second quadrant. As this means the final two digits are given by 50-n, and since 50 – 13 = 37, the last two digits are 37 and the final answer is thus 79637.

DO YOU UNDERSTAND?

1) How many digits are there in the square root of 3940575076?

2) What are the first two digits in the square root of 49736582289?

3) The square root of 14153622961 is in the third quadrant. What are the last two digits of this root?

ANSWERS TO THE TIMED TEST PROBLEMS

1) 97 (2 pts)

2) 264 (3 pts)

3) 1245 (5 pts)

4) 4471 (6 pts)

5) 14615 (7 pts)

6) 56069 (8 pts)

7) 118153 (9 pts)

8) 302751 (10 pts)

ANSWERS TO 'DO YOU UNDERSTAND?'

1) The root has five digits.

2) The first two digits are 22.

3) The last two digits are 69.

Keep adding up those scores, you've three more sections to go.

21

QUICK CURRENCY CONVERSIONS

If you've struggled with changing your pounds into Euros or dollars, then this section is just what you need. When dealing with currency-conversion rates, the rates are presented as 'one-and-four' numbers. Multiplying these would normally give a 'one-and-eight' solution, but only four decimal places are required. Please remember the rules regarding rounding of numbers; if the first digit *beyond* the required accuracy is 5 or higher, you should round your answer *up*. If it is 4 or lower, you should round it *down*.

Calculating exact amounts in different currencies, as opposed to the conversion rates, only needs two decimal places. Again, please remember the 'rounding' rules.

Here are some currency conversion problems for you to try.

Time limit: 18 minutes.

Conversion rates:
$1 = A$1.7542
€1 = $1.1386
£1 = €1.3944

Give each of the following to four decimal places:
1) €1 in A$ (5 pts)
2) £1 in $ (5 pts)
3) £1 in A$ (6 pts)

Give each of the following to two decimal places:
4) €5.73 in A$ (6 pts)
5) A$6.29 in € (6 pts)
6) £5.34 in $ (6 pts)
7) $7.22 in £ (6 pts)
8) A$17.49 in £ (8 pts)

This section can be dealt with quickly, since most of the work has already been covered in the decimal multiplications and divisions sections.

When multiplying two conversion rates, such

as $/A$ and €/$ as above, this is the equivalent of a one-and-four by one-and-four problem from decimal multiplications. However, the size of the answer is one-and-eight, but the problem calls for a one-and-four answer. The solution is clear; simply don't write down the first four digits you create. However, you should remember what these digits are for future reference, and you must also remember the effect of rounding when you do start recording the answer. For instance, 1.7392 x 1.1412 = 1.98477504 which, to four decimal places, is 1.984<u>8</u> and not 1.984<u>7</u>.

When you multiply a given conversion rate by one which you have calculated, this is when you must remember by how much you rounded up or down the rate you calculated. This amount must then be used to temper your answer to the latest problem. For instance, the answer given to the problem above has been rounded up by approximately 0.25 in the last digit. If you now multiply that answer (as written) by 1.3865, the answer to 8 decimal places is 2.75192520. This must now be tempered by the effect of rounding. Having created the answer of 1.9848 by rounding up by 0.25 in the last digit, and with this answer being multiplied by 1.3865, the current answer should be rounded down by (0.25 x 1.3865 = 0.3466 approx.) in the fourth digit, ie, 0.00003466, thus giving 2.75189054. This, when given to four decimal places, is 2.7519, which is the correct answer.

With conversion rates accurate to four places, it is very unlikely that any such tempering would be required when calculating amounts to two decimal places.

CHAMPION'S TOP TIP

This is relatively simple. Remember the tips and techniques regarding multiplications? Of course you do. Just apply those same methods and remember to keep a mental note of your 'shortfall' or 'excess' when you round your answer to 4 decimal places (or however many you are asked for).

ANSWERS TO THE TIMED TEST PROBLEMS

1) €1 = A\$1.9973 (5 pts)
2) £1 = \$1.5877 (5 pts)
3) £1 = A\$2.7851 (6 pts)
4) €5.73 = A\$11.44 (6 pts)
5) A\$6.29 = €3.15 (6 pts)
6) £5.34 = \$8.48 (6 pts)
7) \$7.22 = £4.55 (6 pts)
8) A\$17.49 = £6.28 (8 pts)

Keep totting up those scores, just two more sections left before you face the test set in the style of the Mental Calculations World Championships...

22

THE RAPID WAY TO DEEP ROOTS

The term 'deep roots' is used here to describe a 'root' that is deeper than the square root of a number. For instance, the cube root is the number which must be multiplied three times over (i,e. N x N x N = original number), the fourth root must be multiplied four times (N x N x N x N) and so on. To make things even more difficult, these problems almost *never* have exact answers; you need to find each 'deep root' as a number which is accurate to two decimal places. Good luck, here goes...

Here are three very knotty little problems for you to try.

Time limit: 18 minutes.

Give the answers to the following problems to two decimal places:

1) Find the cube root of 29000 (5 pts)
2) Find the fourth root of 60000 (7 pts)
3) Find the fifth root of 93000 (10 pts)

As you've probably guessed, this is not an easy group of problems to handle. Some use of differential calculus would be useful for this section, and it can help to find the answers not only to cube roots, but also deeper roots as well. Here's how it works.

A line drawn on a graph may be represented by the equation $y = an^b$, where y represents the total value (equivalent to the original number), a is called the coefficient, n is the equivalent of the root, and b is known as the index. Making a first estimate of a deep root and substituting this for n, it is possible to see how accurate or inaccurate this estimate is by applying the relevant number of multiplications. The equation for calculating the gradient (slope) of the line is found by multiplying the coefficient by the index and reducing the index by 1. Performing this adjustment to the equation again will give us the rate of change of the gradient.

HERE'S THE PROBLEM: FIND THE FOURTH ROOT OF 15000 TO 2 DECIMAL PLACES
THIS IS WHAT YOU DO:

- The number 15000 has a fourth root between 11 and 12, since 11^4 (11 x 11 x 11 x 11) = 14641, and 12^4 (12 x 12 x 12 x 12) = 20736. The first estimate of 11 is much closer, with the result of 14641 (15000 − 14641 = 359) being too low.

- The original estimate of 11 needs to be increased, and we need to check the gradient and the rate of change in order to find out how much of an increase is required.

- Since the index is 4 and the coefficient is presently 1, the gradient at the point where n = 11 (the first estimate) is found by taking the new coefficient as 1 x 4 = 4 and the new index as 4 − 1 = 3. This gives the gradient as 4×11^3 (4 x 11 x 11 x 11) = 5324. The rate of change can be found by taking a coefficient of 4 x 3 = 12 and an index of 3 − 1 = 2. This leaves us with a rate of change of 12×11^2 (12 x 11 x 11) = 1452.

- Temporarily disregarding the rate of change, we can make an estimate of the required increase by dividing the shortfall (359) by the gradient (5324), resulting in an estimated increase of 359 / 5324 = 0.07 to two decimal places. Since the rate of change is 1452, the gradient at n = 11.07 would be approximately 0.07 x 1452 (101.64, around 102) higher than at n = 11. Since 5324 + 102 = 5426, the gradient at n = 11.07 is 5426. Assuming the gradient will change at a uniform rate, the average gradient between n = 11 and n = 11.07 will be (5324 + 5426) / 2 = 5375. Applying this average gradient to the increase in the value of n (0.07), we find the increase in y (the total value) is 5375 x 0.07 = 376.25. As we needed to increase the value of y by the shortfall of 359, we find the total is now 376.25 - 359 = 17.25, too high.

- As our new estimate of n = 11.07 is much closer than our original estimate of n = 11, we can now disregard the effect of the rate of change of gradient. We have already found that the gradient of the line at n = 11.07 is 5426, and so we divide the excess value (17.25) by the gradient (5426) to determine the amount by which we need to reduce our present estimate of n = 11.07. Since 17.25/5426 = 0.0032 to four decimal

places, and 11.07 - 0.0032 = 11.0668, the
fourth root is now seen to be 11.0668 to four
decimal places. Since the problem asks for
an answer to only two decimal places, the
answer should be given as 11.07.

- To show how accurate these estimates can
get, the fourth root of 15000 to four decimal
places actually *is* 11.0068.

Let's see this method in action again.

HERE'S THE PROBLEM: FIND THE FIFTH ROOT OF 88000 TO 2 DECIMAL PLACES
THIS IS WHAT YOU DO:

- The fifth root of 88000 is between 9 and 10,
since 9^5 (9 x 9 x 9 x 9 x 9) = 59049 and 10^5
(10 x 10 x 10 x 10 x10) = 100000. Since 10
provides a closer result to the target value of
88000, our first estimate is 10 with an excess
of 100000 – 88000 = 12000.

- The values are: n = 10, coefficient = 1, index
= 5. To find the gradient, we multiply the
coefficient by the index (1 x 5 = 5) and reduce
the index by 1 (5 – 1 = 4). The value of
5×10^4 (5 x 10 x 10 x 10 x 10) = 50000 and this

is therefore the gradient. If the new coefficient is multiplied by the new index (5 x 4 = 20) and the index reduced again (4 − 1 = 3), the rate of change of gradient is given by 20 x 10^3 (20 x 10 x 10 x 10) and is thus 20000.

- Again we temporarily disregard the effects of the rate of change and focus on the gradient to find the first adjustment to the value of n, currently standing at 10. Dividing the excess by the gradient, we find 12000 / 50000 = 0.24 and so this is the first adjustment. Consequently, the new estimate is 10 − 0.24 = 9.76.

- Since the gradient at n = 10 is 50000, the rate of change is 20000 and the reduction of the estimate is 0.24, the gradient is reduced by 0.24 x 20000 = 4800 to become 50000 − 4800 = 45200. The average gradient between n = 10 and n = 9.76 is therefore (50000 + 45200) / 2 = 47600, and therefore the reduction in overall value is thus 47600 x 0.24 = 11424. This means the new estimate of 9.76 give a value for y = n^5 of 100000 − 11424 = 88576, which is still 88576 − 88000 = 576, too high.

- Using the calculated gradient of 45200 at n = 9.76, the excess value of n is approximately 576 / 45200 = 0.0127 to four decimal places leaving a final figure of 9.76 – 0.0127 = 9.7473 to four decimal places and therefore 9.75 to two places. This is again the correct answer, and the accuracy of this method is again demonstrated by the fact that the actual fifth root of 88000 to 4 decimal places is 9.7476, just 0.0003 away from the estimate.

DO YOU UNDERSTAND?

1) What is the best 'first estimate' for the fourth root of 29000?

2) What is the gradient at n = 8 when calculating a fifth root?

3) What is the rate of change at the same point as in question 2?

CHAMPION'S TOP TIP

When finding a deep root, don't just find the nearest whole number to begin with – also seek the result from this number and the next one up or down. This will give you a good indication of how much you are likely to need to add or subtract from your 'basic' number to find the correct answer.

ANSWERS TO THE TIMED TEST PROBLEMS

1) 30.72 (5 pts)
2) 15.65 (7 pts)
3) 9.86 (10 pts)

ANSWERS TO 'DO YOU UNDERSTAND?'

1) The best is 13, since $13^4 = 28561$.
2) The gradient is $5 \times 8^4 = 20480$.
3) The rate of change is $20 \times 8^3 = 10240$.

DAYS OF THE WEEK

These problems, if not particularly easy, are actually quite straightforward. The calculation system relies on all dates being worked out through the Gregorian calender system. The task itself is simple enough: here is a date from the past or future, what day of the week does it represent?

Here are three problems for you to solve.

Time limit: 5 minutes.

On what day of the week was or will be each of the following dates?
1) June 26, 1982 (5 pts)
2) November 6, 1994 (5 pts)
3) July 2, 2014 (5 pts)

- The first step is to choose the starting point for the calculations. Start from January 1, 2001 (a Monday) for all dates after this date, or from January 1, 1901 (a Tuesday) for all dates up to December 31, 2000. If your starting point is later than your target date, or if it is 100 years or more earlier than the target, move forwards (or back) century by century from your starting point until you reach a date within 100 years *earlier* than the target date. Starting at zero, keep a running total of numbers – and add or subtract 5 for each century (if any) you move forwards or backwards and an extra 1 for each fourth century.

- Next, move forwards by as many jumps of 28 years as you can without going past the target date. No adjustment is to be made to the running total for these moves.

- The next step is to move forward as many jumps of four years as you can, again without passing the target date. For each jump, add 5 to your running total.

- If the current date is not yet in the same year as the target date, you should now move forwards as many years as you need to in order to reach the year of the target date. Each one-year jump adds 1 to the running total.

- Once the current date is in the same year as the target date, the next step is to move forwards as many months as is necessary in order to reach the month of the target. For the running total, add the number of days in each month you move out of (i.e;. add 31 for moving from January to February or 30 for moving from April to May).

- Since it is important to remember that February has 29 days rather than its usual 28 days if the year is a leap year, it is therefore necessary to know if the year is a leap year. Here are the rules for this:

- If the year number is divisible by 400 then it is a leap year.

- If the year number is divisible by 4 but not by 100 then it is a leap year.

- If neither of the above is true, then it is not a leap year.

To make the movement between months a little simpler, you can move quarter by quarter as well. Moving directly to April 1 (second quarter) subtracts 1 from the running total unless it is a leap year, in which case there is no adjustment. No adjustment is made for moving from here to quarter 3 (July 1) and moving from quarter 3 into quarter 4 (October 1) adds 1 to the running total.

Once you have reached the month of the target date, add the date number to the running total and subtract 1. If the running total is negative, or if it is 7 or more, add or subtract 7 as many times as is necessary to bring the number into the range 0-6. The final step is to move forwards this number of days from your starting point. The day you reach with this step will be the day of the week of the target date.

HERE'S THE PROBLEM: ON WHAT DAY OF THE WEEK WAS SEPTEMBER 19, 1846?
THIS IS WHAT YOU DO:

- As this date is prior to January 1, 2001, the starting point for the calculations is Tuesday, January 1, 1901.
- Move back one century (to January 1, 1801) and subtract 5 from the running total leaving a total so far of -5.
- Now move forwards by one jump of 28 years (to January 1, 1829) and make no

adjustment to the running total.

- Move forwards by four jumps of 4 years (to January 1, 1845) and add 5 to the running total for each of these jumps. Since 4 x 5 = 20 and -5 + 20 = 15, the total now stands at 15.

- Move forwards 1 year (to January 1, 1846) and add 1 to the running total, thus making it 16.

- Now move forwards to the correct month. Since 1846 was not a leap year, moving to quarter 2 causes the running total to be reduced by 1. Moving to quarter 3 has no effect on the total, moving from July to August adds 31, and moving from August to September also adds 31. The total now stands at 77.

- Finally, add the date number (19) to the running total (77) and subtract 1. The running total is now 95, and is lowered by 91 (13 x 7) to leave 4. Moving forwards by 4 days from the day of the starting point (Tuesday) we reach Saturday which is the right answer.

HERE'S THE PROBLEM: ON WHAT DAY OF THE WEEK WAS MARCH 14, 1964?

THIS IS WHAT YOU DO:

- As this date is prior to January 1, 2001, the starting point for the calculations is Tuesday, January 1, 1901.
- As the starting point is already in the correct century, no movements between centuries are required. The running total thus still stands at 0.
- Move forwards by two jumps of 28 years to January 1, 1957. This change has no effect on the running total which is still 0.
- Next, move forwards by one 4-year jump to January 1, 1961 and add 5 to the running total, thus making 5.
- Now move forwards by 3 single years to January 1, 1964 and add 1 to the running total for each year. The total so far is thus 8.
- It's now time to move to the correct month: moving from January to February adds 31 to the running total, making 39, and the move from February to March adds 29 since 1964 was a leap year. This leaves a running total of 68 so far.
- Adding the date number (14) to the running

total (68) and subtracting 1, the total becomes 81. Subtracting 77 (11 x 7) from this total leaves 4, and moving forwards 4 days from Tuesday (from the starting point) we arrive at a Saturday. Consequently, Saturday is the answer to be presented.

HERE'S THE PROBLEM: ON WHAT DAY OF THE WEEK WILL DECEMBER 5, 2063 FALL?
THIS IS WHAT YOU DO:

- As this date is later than December 31, 2000 the starting point is Monday, January 1 2001. No movement between centuries is required.
- Next move forwards by two jumps of 28 years to January 1, 2057. This change has no effect on the running total which is still 0.
- Now move forwards by one 4-year jump to January 1, 2061 and add 5 to the running total, thus making 5.
- Move forwards by two single years to January 1, 2063 and add 1 to the running total for each year. The total so far is thus 7.
- The time has come again to move forwards to the correct month. Moving to the second quarter (April 1) subtracts 1 from the total,

since 2063 will not be a leap year. Moving to the third quarter (July 1) does not change the total, and moving to the fourth quarter (October 1) adds 1. The move from October to November adds 31, and the move from November to December adds 30. The running total is now 68.

- Adding the date number (5) to the total (68) and subtracting 1, the total becomes 72. Subtracting 70 (10 x 7) leaves 2, and moving forwards 2 days from Monday (from the starting point) we arrive at Wednesday, which is correct.

CHAMPION'S TOP TIP

After finding several answers, it is a simple matter to see if they 'fit' together. By adding the number of years of difference between one date and another, then the number of leap years, and also by taking account of the number of days in each month, you can see if one day will naturally lead to another. If one date 'falls out of line', you have got something wrong and it is time to check again. If they all work together, this is a strong indication that they are all correct. You might even try linking one date, in much the same manner, to a date you know – such as the day you were born.

ANSWERS TO THE TIMED TEST PROBLEMS

1) Saturday (5 pts)
2) Sunday (5 pts)
3) Wednesday (5 pts)

Well, that's the last section dealt with – how did you do? Add up all the points you scored while working through this book and compare your total with the table below.

0-79 points	Poor
80-239 points	Fair
240-439 points	Good
440-599 points	Excellent
600-674 points	Consider entering the Mental Calculations World Championship
675-709 points	Potential medallist at the World Championships
710-750 points	Potential World Champion

You are now be ready to take on a major test. What follows is a fourteen-page test equivalent to the Mental Calculations World Championship (the answers are given on six further pages). You will face a total of 140 problems, and you have a time limit of four hours. Please do not start this test unless and until you are sure you can commit four hours of your time. You should make sure that you are happy with the methods discussed in this book. If there are any sections you are not sure of, *read through those sections again before taking the test.*

The rules for the test are as follows:

1) Do not obtain help from another person while you are taking the test.
2) Do not look at the questions before you start the test.
3) Do not look at the answers before you have finished.
4) Do not use any calculating aids.
5) Do not write down any intermediate calculations.
6) If you feel the need to do so, you may change each answer *once only.*
7) You score the number of points given if your answer is correct. If your answer is wrong, you score zero points – however close your answer is. (Remember that trailing zeros *after* a decimal point need not be recorded unless they have been specifically requested – the number 23.2500 is *exactly the same* as 23.25.)

8) Do not take longer than the allotted time limit of four hours.

9) Once you have finished the test, check your answers and add up your score. Compare this with both the table above and your total from the sections in this book; hopefully your score will have improved.

Good luck, maybe we'll compete against each other at a future event.

24

CHAMPIONSHIP QUIZ

Now have a go at the the kind of quiz that faces the most skilled champions in the maths world. Are you ready for it yet? Answers follow...

INTEGER ADDITIONS

GEORGE LANE

Problem					Answer
16 +	53				1
67 +	42				1
361 +	468 +	874			2
836 +	623 +	736			2
94997 +	8222 +	92587			3
747177 +	51215 +	230793 +	37720		5
9948945 +	549171 +	7019601 +	940756		6
3207355 +	933928 +	2980641 +	577828 +	4427994	8
86655851 +	5680989 +	84179578 +	1939108 +	18140215	10
					38

INTEGER SUBTRACTIONS

171	-	77
148	-	83
12278	-	3721
142513	-	75109
1110580	-	237922
111584852	-	21601167
142422258	-	81778129
8550978227	-	4344446043

Boxes: 1 1 2 3 4 6 8 10 35

INTEGER MULTIPLICATIONS

4	x	82
58	x	59
148	x	7899
4181	x	4857

Boxes: 1 2 4 5

7
8
9
10
46

1
2
4
5
6
8
9
10
45

10359 × 52168
15651 × 198566
201999 × 179687
207581 × 2758358

INTEGER DIVISIONS

Each problem has an exact integer solution.

468 / 6
588 / 12
214848 / 373
2328524 / 484
28345418 / 3994
1175655756 / 88183
3131783072 / 14972
57516133392 / 227412

DECIMAL ADDITIONS

1   5.5 + 5.4

2   16.9 + 16.1

2   38.7 + 10.9

4   872.73 + 4859.22 + 5324.177

5   6846.595 + 60436.148 + 9243.092

6   33101.066 + 14591.575 + 22353.942

7   37228.2445 + 47609.5076 + 33488.4169

8   26664.4045 + 413601.6788 + 70136.3868 + 836067.9571

10   3587578.2798 + 888685.45226 + 421896.53534 + 570021.02125 + 8508229.00135

45

DECIMAL SUBTRACTIONS

1   147.6 - 98.1

2   1581.9 - 861.5

3   1125.15 - 541.99

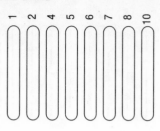

3   4   6   8   10   37

1   2   4   5   6   7   8   10   43

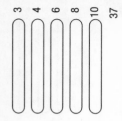

3665.67 -	1381.68
5599.919 -	2674.76
145060.847 -	73630.214
270420.7074 -	124812.0878
1521562.25817 -	414221.83223

DECIMAL MULTIPLICATIONS

3.6	×	9.8
73.7	×	67.4
446.8	×	814.2
91.12	×	152.89
781.35	×	379.76
351.59	×	4096.24
7267.28	×	2894.93
9735.013	×	1465.273

DECIMAL DIVISIONS

Each problem has an exact solution.

61.74	/	6.3
383.37	/	3.9
2300.61	/	34.7
75914.88	/	182.4
47274.054	/	529.8
1437594.1812	/	6236.31
10433442.7022	/	5810.14
354728.663025	/	731.775

1
2
3
5
6
8
9
10

43

FRACTION PROBLEMS

Give the answer to each fraction problem as a mixed number in its simplest form,

e.g.; '13 1/3 x 3 1/12 = 41 1/9, not 370/9 or 41 4/36'.

FRACTION ADDITIONS

8 11/20 +   11 5/13

7 5/6 +   1 7/18

15 3/4 +   2 1/2

6 5/6 +   8 2/3

9 1/2 +   15 3/7

19 1/3 +   18 8/15

13 17/19 +   2 2/7

1 3/8 +   19 1/10

15 7/8 +   1 1/3

GEORGE LANE

4
4
4
4
4
4
4
4
4

36

FRACTION SUBTRACTIONS

$18\ \frac{3}{5}$	-	$11\ \frac{3}{4}$		5
$21\ \frac{3}{4}$	-	$8\ \frac{3}{14}$		5
$6\ \frac{11}{19}$	-	$5\ \frac{11}{13}$		5
$26\ \frac{1}{2}$	-	$7\ \frac{11}{18}$		5
$22\ \frac{1}{2}$	-	$5\ \frac{1}{8}$		5
$29\ \frac{2}{3}$	-	$17\ \frac{1}{13}$		5
$21\ \frac{3}{4}$	-	$19\ \frac{4}{5}$		5
$19\ \frac{3}{20}$	-	$4\ \frac{5}{6}$		5
$21\ \frac{1}{2}$	-	$11\ \frac{5}{6}$		5
				45

FRACTION MULTIPLICATIONS

$17\ \frac{11}{20}$	x	$10\ \frac{1}{6}$		5
$4\ \frac{2}{3}$	x	$5\ \frac{3}{5}$		5
$12\ \frac{8}{9}$	x	$12\ \frac{3}{17}$		5

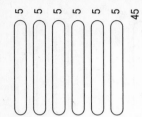

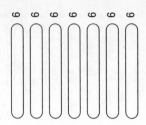

$2\frac{2}{3}$ x	$16\frac{3}{7}$		5
$6\frac{1}{2}$ x	$1\frac{10}{17}$		5
$7\frac{5}{8}$ x	$19\frac{11}{17}$		5
$6\frac{1}{2}$ x	$1\frac{1}{4}$		5
$5\frac{3}{5}$ x	$15\frac{9}{11}$		5
$10\frac{2}{5}$ x	$1\frac{2}{7}$		5
			45

FRACTION DIVISIONS

$6\frac{1}{2}$ /	$19\frac{1}{3}$		6
$18\frac{1}{2}$ /	$13\frac{1}{3}$		6
$11\frac{11}{20}$ /	$16\frac{1}{4}$		6
$18\frac{3}{10}$ /	$12\frac{5}{12}$		6
$5\frac{13}{17}$ /	$6\frac{1}{8}$		6
$6\frac{2}{3}$ /	$13\frac{2}{5}$		6
$15\frac{12}{19}$ /	$10\frac{2}{3}$		6

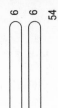

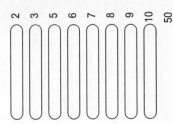

6
6
54

2
3
5
6
7
8
9
10
50

10 ⅓ /   7 ⅗₁
9 ⅜ /   4 ⅝₀

SQUARE ROOTS

Each problem has an exact integer solution.

1156
34225
6584356
64754209
780867136
5347119376
41064590736
91843545249

213

REMAINDERS

Find the remainder after an inexact division, e.g.; '461 / 52 = 8 with remainder 45, so the answer is 45'.

174	/	38
94987	/	897
146733	/	750
624664	/	6310
1604709	/	9470
5355113	/	3412
254079366	/	8455
921390458	/	46588

GEORGE LANE

2
4
5
6
7
8
9
10

51

PRIME FACTORS

Give the prime numbers which multiply together to give the required number,

e.g.; '420 = 2 x 2 x 3 x 5 x 7, so the answer would be 2, 2, 3, 5, 7'.

174

1564

7038

33235

66470

340170

917286

1870544

2

4

5

6

7

8

9

10

51

DEEP ROOTS

Give the required answer to 2 decimal places.

Cube root of	36000	☐
Fourth root of	73000	☐
Fifth root of	83000	☐

5
7
10
22

CURRENCY CONVERSION

Conversion rates

$1 = A$1.7984

€1 = $1.1734

£1 = €1.4126

Give each of the following to four decimal places.

€1 in A$	☐
£1 in $	☐
£1 in A$	☐

5
5
6

216

Give each of the following to two decimal places.

€6.37 in A$ 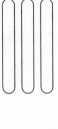 6

A$7.91 in E   6

£4.83 in $ 6

$8.64 in £   6

A$23.77 in £   8

                               48

DAYS OF THE WEEK

On what day of the week is or will be each of the following dates?

October 5, 1957   5

August 28, 1988   5

November 27, 2014   5

                               15

217

TOTAL 750

25

CHAMPIONSHIP QUIZ
ANSWERS

Well, now you have a reasonable idea of how it feels to take part in the annual number-skills test we call the Mental Calculations World Championship. The following six pages carry the questions again, this time with the answers printed in the boxes. Remember to score points only for correct answers; you score zero for a wrong answer regardless of how close you may be. Remember also that, although trailing zeros are not generally required after a decimal point, it should not be regarded as wrong to include them. If the answer to a problem is required to have a specific number of decimal places, this number must be used – no fewer and no more.

INTEGER ADDITIONS

PROBLEM	SOLUTION	POINT
16 + 53	69	1
67 + 42	109	1
361 + 468 + 874	1703	2
836 + 623 + 736	2195	2
94997 + 8222 + 92587	195806	3
747177 + 51215 + 230793 + 37720	1066905	5
9948945 + 549171 + 7019601 + 940756	18458473	6
3207355 + 933928 + 2980641 + 577828 + 4427994	12127746	8
86655851 + 5680989 + 84179578 + 1939108 + 18140215	196595741	10
		38

1	94
1	65
2	8557
3	67404
4	872658
6	89983685
8	60644129
10	4206532184
35	

1	328
2	3422
4	1169052
5	20307117

INTEGER SUBTRACTIONS

```
     171 -         77
     148 -         83
   12278 -       3721
  142513 -      75109
 1110580 -     237922
111584852 -   21601167
142422258 -   81778129
8550978227 - 4344446043
```

INTEGER MULTIPLICATIONS

```
   4 x   82
  58 x   59
 148 x 7899
4181 x 4857
```

10359 ×	52168
15651 ×	198566
201999 ×	179687
207581 ×	2758358

540408312	7
3107756466	8
36296594313	9
572582711998	10
	46

78	1
49	2
576	4
4811	5
7097	6
13332	8
209176	9
252916	10
	45

INTEGER DIVISIONS

Each problem has an exact integer solution.

468 /	6
588 /	12
214848 /	373
2328524 /	484
28345418 /	3994
1175655756 /	88183
3131783072 /	14972
57516133392 /	227412

DECIMAL ADDITIONS

5.5 +	5.4				10.9
16.9 +	16.1				33
38.7 +	10.9				49.6
872.73 +	4859.22	5324.177			11056.127
67846.595 +	60436.148	9243.092			137525.825
33101.066 +	14591.575 +	22353.942			102928.502
37228.2445 +	47609.5076 +	33488.4169			150320.311
26664.4045 +	413601.6788 +	836067.9571			1346470.4272
3587578.2798 +	888685.45226 +	421896.53534 +	570021.02125 +	8508229.00135	13976410.29

1
2
2
4
5
6
7
8
10
45

DECIMAL SUBTRACTIONS

147.6 -	98.1		49.5
1581.9 -	861.5		720.4
1125.15 -	541.99		583.16

1
2
3

222

3665.67 − 1381.68   (2283.99)   3
5599.919 − 2674.76   (2925.159)   4
145060.847 − 73630.214   (71430.633)   6
270420.7074 − 124812.0878   (127608.6196)   8
1521562.25817 − 414221.83223   (11073340.42594)   10
     37

DECIMAL MULTIPLICATIONS

3.6 × 9.8   (35.28)   1
73.7 × 67.4   (4967.38)   2
446.8 × 814.2   (363784.56)   4
91.12 × 152.89   (13931.3368)   5
781.35 × 379.76   (296725.476)   6
351.59 × 4096.24   (1440197.0216)   7
7267.28 × 2894.93   (21038266.8904)   8
9735.013 × 1465.273   (14264451.708549)   10
     43

DECIMAL DIVISIONS

Each problem has an exact solution.

61.74 /	6.3	9.8	1
383.37 /	3.9	98.3	2
2300.61 /	34.7	66.3	3
75914.88 /	182.4	416.2	5
47274.054 /	529.8	89.23	6
1437594.1812 /	6236.31	230.52	8
10433442.7022 /	5810.14	1795.73	9
354728.663025 /	731.775	484.751	9
			43

224

FRACTION PROBLEMS

Give the answer to each fraction problem as a mixed number in its simplest form,

e.g.; '$13\frac{1}{3} \times 3\frac{1}{12} = 41\frac{1}{9}$, not $\frac{370}{9}$ or $41\frac{4}{36}$'

FRACTION ADDITIONS

$8\frac{11}{20}$ +	$11\frac{5}{13}$	$19\frac{243}{260}$	4
$7\frac{5}{8}$ +	$1\frac{7}{18}$	$9\frac{1}{72}$	4
$15\frac{3}{5}$ +	$2\frac{1}{2}$	$18\frac{1}{10}$	4
$6\frac{2}{3}$ +	$8\frac{2}{13}$	$14\frac{32}{39}$	4
$9\frac{1}{2}$ +	$15\frac{3}{7}$	$24\frac{13}{14}$	4
$19\frac{1}{3}$ +	$18\frac{8}{15}$	$37\frac{13}{15}$	4
$13\frac{17}{19}$ +	$2\frac{4}{9}$	$16\frac{58}{171}$	4
$1\frac{5}{6}$ +	$19\frac{1}{10}$	$20\frac{14}{15}$	4
$15\frac{7}{8}$ +	$1\frac{1}{5}$	$17\frac{3}{40}$	4
			36

FRACTION SUBTRACTIONS

$18 \frac{3}{5}$ -	$11 \frac{3}{4}$		$6 \frac{17}{20}$	5
$21 \frac{3}{4}$ -	$8 \frac{3}{14}$		$13 \frac{15}{28}$	5
$6 \frac{11}{19}$ -	$5 \frac{11}{13}$		$\frac{181}{247}$	5
$26 \frac{1}{2}$ -	$7 \frac{11}{18}$		$18 \frac{8}{9}$	5
$22 \frac{1}{2}$ -	$5 \frac{1}{8}$		$17 \frac{3}{8}$	5
$29 \frac{2}{3}$ -	$17 \frac{1}{13}$		$12 \frac{23}{39}$	5
$21 \frac{3}{4}$ -	$19 \frac{4}{5}$		$1 \frac{19}{20}$	5
$19 \frac{3}{20}$ -	$4 \frac{5}{6}$		$14 \frac{19}{60}$	5
$21 \frac{1}{2}$ -	$11 \frac{5}{6}$		$9 \frac{2}{3}$	5
				45

FRACTION MULTIPLICATIONS

$17 \frac{1}{20}$ x	$10 \frac{1}{6}$		$178 \frac{17}{40}$	5
$4 \frac{2}{3}$ x	$5 \frac{3}{5}$		$26 \frac{2}{15}$	5
$12 \frac{8}{9}$ x	$12 \frac{3}{7}$		$156 \frac{16}{17}$	5

$2\frac{2}{3}$ x	$16\frac{3}{7}$	**$43\frac{17}{21}$**	5
$6\frac{1}{2}$ x	$1\frac{10}{17}$	**$10\frac{11}{34}$**	5
$7\frac{5}{8}$ x	$19\frac{11}{17}$	**$149\frac{55}{68}$**	5
$6\frac{1}{2}$ x	$1\frac{1}{4}$	**$8\frac{1}{8}$**	5
$5\frac{3}{5}$ x	$15\frac{9}{11}$	**$88\frac{32}{55}$**	5
$10\frac{2}{5}$ x	$1\frac{2}{7}$	**$13\frac{13}{35}$**	5
			45

FRACTION DIVISIONS

$6\frac{1}{2}$ /	$19\frac{1}{3}$	$\frac{39}{116}$	6
$18\frac{1}{2}$ /	$13\frac{1}{3}$	**$1\frac{31}{80}$**	6
$11\frac{11}{20}$ /	$16\frac{1}{4}$	$\frac{231}{325}$	6
$18\frac{3}{10}$ /	$12\frac{5}{12}$	**$1\frac{353}{745}$**	6
$5\frac{13}{17}$ /	$6\frac{1}{8}$	$\frac{16}{17}$	6
$6\frac{2}{3}$ /	$13\frac{2}{5}$	$\frac{100}{201}$	6
$15\frac{12}{19}$ /	$10\frac{2}{3}$	**$1\frac{283}{608}$**	6

1 $\frac{104}{237}$	6
1 $\frac{47}{49}$	6
	54

34	2
185	3
2566	5
80457	6
27944	7
73124	8
202644	9
303057	10
	50

10 $\frac{1}{3}$ /   7 $\frac{7}{11}$

9 $\frac{3}{5}$ /   4 $\frac{9}{10}$

SQUARE ROOTS

Each problem has an exact integer solution.

1156

34225

6584356

64754209

780867136

5347119376

4106459073 6

9184354524 9

REMAINDERS

Find the remainder after an inexact division, e.g.; '461 / 52 = 8 with remainder 45, so the answer is 45'.

174 /	38	(22) 2
94987 /	897	(802) 4
146733 /	750	(483) 5
624664 /	6310	(6284) 6
1604709 /	9470	(4279) 7
5355113 /	3412	(1685) 8
254079366 /	8455	(6616) 9
921390458 /	46588	(19582) 10
		51

PRIME FACTORS

Give the prime numbers which multiply together to give the required number,

e.g.; '420 = 2 x 2 x 3 x 5 x 7, so the answer would be 2, 2, 3, 5, 7'.

Number	Answer	#
174	2, 3, 29	2
1564	2, 2, 17, 23	4
7038	2, 3, 3, 17, 23	5
33235	5, 17, 17, 23	6
66470	2, 5, 17, 17, 23	7
340170	2, 3, 5, 17, 23, 29	8
917286	2, 3, 17, 17, 23, 23	9
1870544	2, 2, 2, 2, 13, 17, 23, 23	10

51

DEEP ROOTS

Give the required answer to 2 decimal places.

Cube root of	36000	5
Fourth root of	73000	7
Fifth root of	83000	10

33.02

16.44

9.63

22

CURRENCY CONVERSION

Conversion rates

$1 = A$1.7984

€1 = $1.1734

£1 = €1.4126

Give each of the following to four decimal places.

€1 in A$	5
£1 in $	5
£1 in A$	6

2.1102

1.6575

2.9809

Give each of the following to two decimal places.

€ 6.37 in A$	(13.44) 6
A$7.91 in €	(3.75) 6
£4.83 in $	(8.01) 6
$8.64 in £	(5.21) 6
A$23.77 in £	(7.97) 8
	48

DAYS OF THE WEEK

On what day of the week is or will be each of the following dates?

October 5, 1957	(Saturday) 5
August 28, 1988	(Sunday) 5
November 27, 2014	(Thursday) 5
	15

TOTAL 750